ANALYSIS AND SYNTHESIS OF DISTRIBUTED REAL-TIME EMBEDDED SYSTEMS

Analysis and Synthesis of Distributed Real-Time Embedded Systems

by

Paul Pop
Linköping University,
Sweden

Petru Eles
Linköping University,
Sweden

and

Zebo Peng
Linköping University,
Sweden

KLUWER ACADEMIC PUBLISHERS
BOSTON / DORDRECHT / LONDON

A C.I.P. Catalogue record for this book is available from the Library of Congress.

ISBN 1-4020-2872-5 (HB)
ISBN 1-4020-2873-3 (e-book)

Published by Kluwer Academic Publishers,
P.O. Box 17, 3300 AA Dordrecht, The Netherlands.

Sold and distributed in North, Central and South America
by Kluwer Academic Publishers,
101 Philip Drive, Norwell, MA 02061, U.S.A.

In all other countries, sold and distributed
by Kluwer Academic Publishers,
P.O. Box 322, 3300 AH Dordrecht, The Netherlands.

Printed on acid-free paper

Printed in the Netherlands.

To our beloved families

Contents

II Time-Driven Systems

III Event-Driven Systems

IV Multi-Cluster Systems

List of Figures

List of Tables

Preface

EMBEDDED COMPUTER SYSTEMS are now everywhere: from alarm clocks to PDAs, from mobile phones to cars, almost all the devices we use are controlled by embedded computers. An important class of embedded computer systems is that of hard real-time systems, which have to fulfill strict timing requirements. As real-time systems become more complex, they are often implemented using distributed heterogeneous architectures.

This book presents analysis and synthesis methods for heterogeneous, distributed hard real-time embedded systems. Such systems are used in many application areas like automotive electronics, real-time multimedia, avionics, medical equipment, and factory systems. These systems are heterogeneous not only in terms of platforms and communication protocols, but also in terms of scheduling approaches.

The proposed analysis and synthesis techniques derive optimized implementations that fulfill the imposed design constraints. An important part of the implementation process is the synthesis of the communication infrastructure, which has a significant impact on the overall system performance and cost.

To reduce the time-to-market of products, the design of real-time systems seldom starts from scratch. Typically, designers start from an already existing system, running certain applications, and the design problem is to implement new functionality on top of this system. Hence, in addition to the analysis and synthesis methods proposed, we also consider mapping and scheduling within such an incremental design process. Supporting incremental design provides a high degree of flexibility, and can result in important reductions of design costs.

The book is structured in four parts, and has ten chapters.

In the first part we present the methodologies commonly used in embedded systems design, and we introduce the application model considered, a control and dataflow graph based representation, called *conditional process graph*. We also present the hardware and software architectures considered, including the details of the communication protocols used: the *time triggered protocol* (TTP), which is a time-driven protocol based on a *time-division multiple access* bus access scheme, and the *controller area network protocol* (CAN), an event-driven communication protocol employing a collision avoidance scheme.

The second part presents analysis and synthesis methods for *time-driven* systems, where processes are activated according to a time-triggered policy. Messages are transmitted using the TTP, while the scheduling of processes is performed using static cyclic scheduling. In such a context, we discuss the static cyclic scheduling of systems with data and control dependencies, and investigate the mapping and scheduling tasks in the context of an incremental design approach.

In the third part we provide an analysis and develop techniques for the synthesis of *event-driven* systems, where the activation of processes is done at the occurrence of significant events. The scheduling of processes is performed using fixed-priority preemptive scheduling. Although a natural mapping of event-driven messages would be on a CAN bus, considering preemptive priority based scheduling at the process level, with time triggered static scheduling at the communication level can be the right solution under several circumstances. Therefore, in part three we will consider that messages produced by event-triggered processes are transmitted using the TTP, and we have developed four message passing policies for transmitting event-triggered messages over a time-triggered bus. Optimization strategies that derive the parameters of the communication protocol are also presented.

The fourth part combines time-driven and event-driven systems into heterogeneous networks, called *multi-cluster* systems. A multi-cluster system consists of several clusters, interconnected by *gateways*. A *cluster* is composed of nodes which share a broadcast communication channel. In a *time-triggered cluster*, processes and messages are scheduled according to a static cyclic policy, with the bus implementing the TTP. On an *event-triggered cluster*, the processes are scheduled according to a priority based preemptive approach, while messages are transmitted using the priority-based CAN protocol. We propose a schedulability analysis technique for such systems, together with optimization strategies for synthesizing a communication infrastructure that guarantees the timing constraints of the application. We also address design problems which are characteristic to multi-clusters: partitioning of the system functionality into time-triggered and event-triggered domains, the mapping of processes to the heterogeneous nodes of a cluster, and the packing of application messages to frames in the case of TTP and CAN protocols.

The approaches presented in the book have been evaluated using a real-life case study consisting of a vehicle cruise controller, and an extensive set of synthetic applications generated for evaluation purposes.

The authors are grateful to Olof Bridal, Magnus Hellring, and Henrik Lönn at Volvo Technology Corporation and Jakob Axelsson at Volvo Car Corporation for their close involvement and valuable feedback during this work. We would like to thank Rolf Ernst from Technical University of Braunschweig, Germany, Hans Hansson from Mälardalen University, Sweden, Axel Jantsch from Royal Institute of Technology, Sweden, and Simin Nadjm-Tehrani from Linköping University, Sweden, for reading an early draft of this book. Many thanks also to Traian Pop and Viacheslav Izosimov for their help with the implementation of some of the algorithms used in the book.

PART I
Preliminaries

Chapter 1
Introduction

THE FIRST MODERN computers occupied entire rooms, had thousands of vacuum tubes, and dissipated hundreds of kilowatts of heat, but could only execute a couple of thousands of simple instructions per second [EB03a]. Today, a complex microprocessor, which dwarfs the performance of the first electronic computers, can be integrated into a digital wristwatch.

This extraordinary development is due to the microelectronics revolution that, according to Moore's law, allows us to double the number of transistors integrated on a single chip every 18 months, from 2,300 in the first Intel 4004 chip to 42 million in the latest Pentium 4 microprocessor [EB03b].

Not only have digital systems become more powerful, and integrated an increasing number of transistors, but their cost has also dropped dramatically. This has led to a situation where we have a huge amount of very cheap computation power available on a very small physical size, allowing the digital systems to be present in every aspect of our daily lives.

Nowadays, digital systems are everywhere. We are surrounded by desktop computers, mobile phones, personal digital assistants (PDAs), pagers, scanners, DVD players, VCRs, video game consoles, fax machines, digital cameras, home security systems, electronic toys, card readers, ATMs, cars, trains, airplanes, etc., all of which contain in some form or another a digital system.

Whenever the digital systems augment or control a function of a host object or system, we say that these digital systems are embedded into the host system, hence the term *embedded system*. Out of the digital systems mentioned above, only the desktop computer is not an embedded system. The desktop computer is a general purpose system that can be programmed to implement virtually any type of function. In contrast, embedded systems are not general purpose systems, rather, their functionality is dedicated to perform a limited set of functions, required by the host system.

Recently, the number of embedded systems in use has become larger than the number of humans on the planet [Aar03]. That is not difficult to believe, considering that more than 99% of the microprocessors produced today are used in embedded systems [Tur99] and the number of microprocessors manufactured has been increasing rapidly over many years. Although the number of embedded systems and their diversity is huge, they share a small, but significant, set of common characteristics:

- Embedded systems are designed to perform a dedicated set of functions.

  For example, a digital camera is designed principally for taking pictures; it cannot be programmed by the user, for example, to solve differential equations.
- Embedded systems have to perform under very tight, varied, and competing constraints.

  A wearable blood pressure monitor has to be small enough to be placed on the wrist, a mobile phone has to consume very little power so the battery lasts for a couple of weeks, and in-vehicle electronics have to perform under tight timing constraints.

- In addition to all these, embedded systems have to be cheap to produce and maintain, and, at the same time, flexible enough to be extended with new dedicated functions whenever necessary.

Therefore, in order to function correctly, an embedded system not only has to be designed such that it implements the required functionality, but also has to fulfill a wide range of competing constraints: development cost, unit cost, size, performance, power consumption, flexibility, time-to-prototype, time-to-market, maintainability, correctness, safety, etc. [Vah02].

As the design of such systems is becoming increasingly complex, new analysis and synthesis techniques are needed. This book presents several analysis and synthesis methods for distributed real-time embedded systems that have recently been developed.

1.1 A Typical Application Area: Automotive Electronics

Although the techniques presented in this book can be successfully used in several application areas, it is useful, for understanding the embedded systems evolution and design challenges, to exemplify the developments in a particular area.

If we take the example of automotive manufacturers, they were reluctant, until recently, to use computer controlled functions onboard vehicles. Today, this attitude has changed for several reasons. First, there is a constant market demand for increased vehicle performance, more functionality, less fuel consumption and less exhausts, all of these at lower costs. Then, from the manufacturers side, there is a need for shorter time-to-market and reduced development and manufacturing costs. These, combined with the advancements of semiconductor technology, which is delivering ever increasing performance at lower and lower costs, has led to the rapid increase in the number of electronically controlled functions onboard a vehicle [Kop99].

The amount of electronic content in an average car in 1977 had a cost of $110. Currently, that cost is $1341, and it is expected that this figure will reach $1476 by the year 2005, continuing to increase because of the introduction of sophisticated electronics found until now only in high-end cars (see Figure 1.1) [Han02], [Lee02]. It is estimated that in 2006 the electronics inside a car will amount to 25% of the total cost of the vehicle (35% for the high end models), a quarter of which will be due to semiconductors [Jos01], [Han02]. High-end vehicles currently have up to 100 microprocessors implementing and controlling various parts of their functionality. The total market for semiconductors in vehicles is predicted to grow from $8.9 billions in 1998 to $21 billion in 2005, amounting to 10% of the total worldwide semiconductors market [Han02], [Kop99].

At the same time with the increased complexity, the type of functions implemented by embedded automotive electronics systems has also evolved. Thanks to the semiconductors revolution, in the late 50s, electronic devices became small enough to be installed on board of vehicles. In the 60s the first analog fuel injection system appeared, and in the 70s analog devices for controlling transmission, carburetor, and spark advance timing were developed. The oil crisis of the 70s led to the demand of engine control devices that improved the efficiency of the engine, thus reducing fuel consumption. In this context, the first microprocessor based injection control system appeared in 1976 in the USA. During the 80s, more sophisticated systems began to appear, like electronically controlled braking systems, dashboards, information and navigation systems, air conditioning systems, etc. In the 90s, development and improvement have concentrated in the areas like safety and convenience. Today, it is not uncommon to have highly critical functions like steering or braking implemented through electronic functionality only, without any mechanical backup, like is the case in drive-by-wire and brake-by-wire systems [Chi96], [XbW98].

A large class of systems have tight performance and reliability constraints. A good example is the engine control unit, whose main task is to reduce the level of exhausts and the fuel consumption by controlling the air and fuel mixture in each cylinder. For this, the engine controller is usually designed as a closed-loop control system which has as feedback the level of exhausts. The engine speed is the most important factor to consider with respect to the timing requirements of the engine controller. A typical 4 cylinder engine has an optimal speed of 6,000 revolutions per minute (RPM). At 6,000 RPM the air to fuel ratio for each cylinder must be recomputed every 20 milliseconds (ms). This means that in a 4 cylinder engine a single controller of

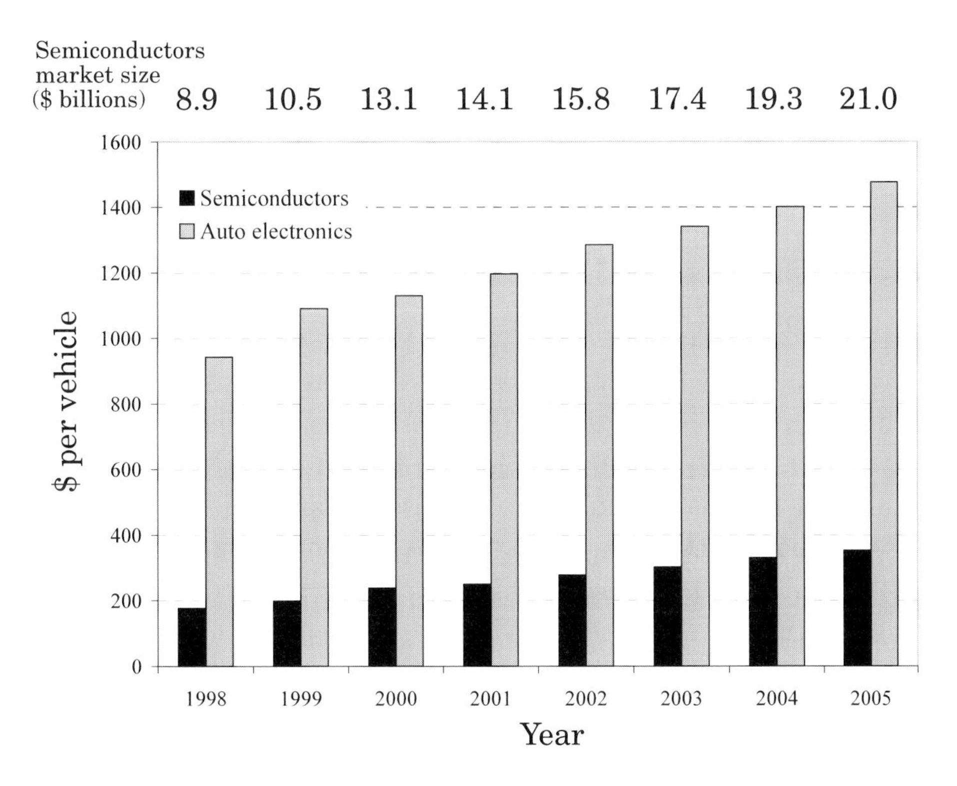

Figure 1.1: Worldwide Automotive Electronics Trends [Han02]

such type must complete the entire loop in 5 ms! For such an engine controller, not meeting the timing constraint leads to a less efficient fuel consumption and more exhausts [Chi96]. However, for other types of systems, like drive-by-wire or brake-by-wire, not fulfilling the timing requirements can have catastrophic consequences.

We have seen so far that the use of electronics in modern vehicles is increasing, replacing or augmenting critical mechanical and hydraulic vehicle components. Their complexity is growing at a very high pace, and the constraints—in terms of functionality, performance, reliability, cost and time-to-market—are getting tighter. Therefore, the task of designing such systems is becoming increasingly important and difficult at the same time. New design techniques are needed, which are able to:

- successfully manage the complexity of embedded systems,
- meet the constraints imposed by the application domain,
- shorten the time-to-market, and
- reduce development and manufacturing costs.

1.2 Distributed Hard Real-Time Embedded Systems

In this book, we are interested in a particular class of systems called *real-time embedded systems*. Very important for the correct functioning of such systems are their timing constraints. For example, a vehicle cruise controller has to react within tens of milliseconds to events originating from the driver or road conditions. Kopetz [Kop97a] gives a definition for a real-time system as being "*a computer system in which the correctness of the system behavior depends not only on the logical results of the computations, but also on the physical instant at which these results are produced.*"

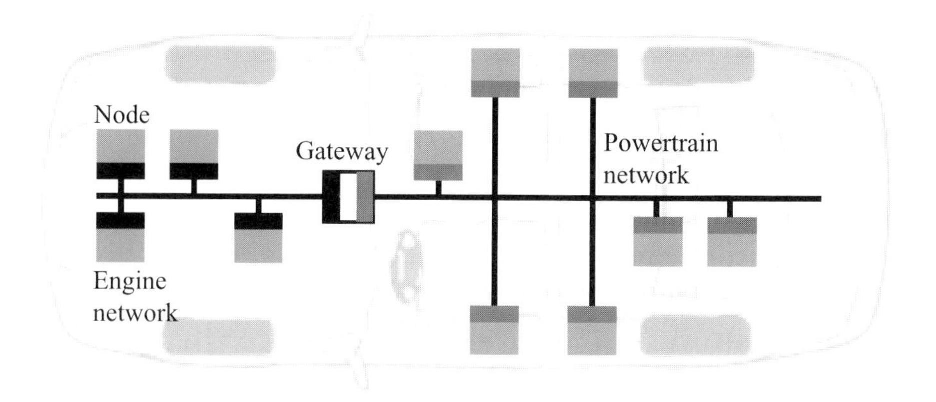

Figure 1.2: A Distributed Real-Time System Example

Real-time systems have been classified as *hard* real-time and *soft* real-time systems [Kop97a]. Basically, hard real-time systems are systems where failing to meet a timing constraint can potentially have catastrophic consequences. For example, a brake-by-wire system in a car failing to react within a given time interval can result in a fatal accident. On the other hand, a multimedia system, which is a soft-real time system, can, under certain circumstances, tolerate a certain amount of delays resulting maybe in a patchier picture, without serious consequences besides some possible inconvenience to the user.

The techniques presented in this book are aimed towards hard-real time systems that implement safety-critical applications where timing constraints are of utmost importance to the correct behavior of the application.

Many such applications, following physical, modularity or safety constraints, are implemented using *distributed architectures*. In Figure 1.2 we illustrate a distributed real-time system implementing some electronic functions of a vehicle. For example, the network on the left is responsible to implement functionality related to the engine, while the network on the right implements functions related to the powertrain, like brake-by-wire, anti blocking system, etc.

Such systems are composed of several different types of hardware components (called *nodes*), interconnected in a network. An increasing number of such systems are today implemented as *heterogeneous* distributed systems, which means that they are composed of several networks, interconnected with each other. Each network has its own communication protocol and two such networks communicate via a *gateway* which is a node connected to both of them. This type of architectures are used in increasing numbers in several application areas: networks on chip are heterogeneous, we also see them in, for example, factory systems, and they are very common in vehicles.

For such systems, the communication between the functions implemented on different nodes has an important impact on the

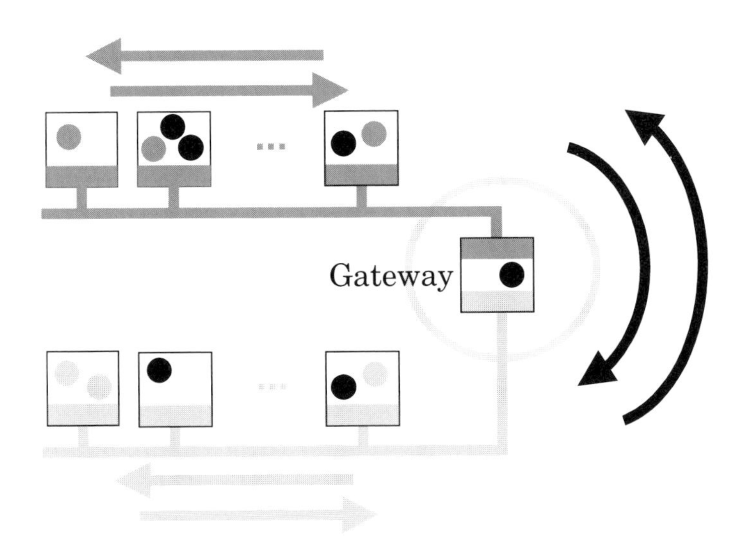

Figure 1.3: Distributed Safety-Critical Applications

overall system properties such as performance, cost, maintainability, etc.

The application software running on such distributed architectures is composed of several functions. The way the functions have been distributed on the architecture has evolved over time. Initially, in automotive applications, each function was running on a dedicated hardware node, allowing the system integrators to purchase nodes implementing required functions from different vendors, and to integrate them together into their system. The number of such nodes in the architecture has exploded, reaching more than 100 in a high-end car. This has created a huge pressure to reduce the number of nodes, use the resources available more efficiently, and thus reduce costs.

This development has led to the need to integrate several functions in one node. For this to be possible, middleware software that abstracts away the hardware differences of the nodes in the heterogeneous architecture has to be available [Eas02]. Using such a middleware architecture, the software functions become independent of the particular hardware details of a node, and thus they can be distributed on the hardware architecture, as depicted in Figure 1.3.

Although an application is typically distributed over one single network, we begin to see applications that are distributed across several networks, like is the case in Figure 1.3 where the third application, represented as black dots, is distributed over the two networks. This trend is driven by the need to further reduce costs, improve resource usage, but also by application constraints like having to be physically close to particular sensors and actuators. Moreover, not only are these applications distributed across networks, but their functions can exchange critical information through the gateway nodes.

Such safety-critical hard real-time distributed applications running on heterogeneous distributed architectures are inherently difficult to analyze and implement. Due to their distributed nature, the communication has to be carefully considered during the analysis and design in order to guarantee that the timing constraints are satisfied under the competing objective of reducing the cost of the implementation.

1.3 Book Overview

We are interested in the analysis and synthesis of safety-critical distributed applications implemented using hard real-time embedded systems.

Several methodologies have been proposed for real-time embedded systems design. Regardless of the chosen methodology, there are a number of major design tasks that have to be performed.

One major design task is to decide what components to include in the hardware architecture and how these components are connected. This is called the *architecture selection* phase. In order to reduce costs, especially in the case of a mass market product, the hardware architecture is usually reused, with some modifications, for several product lines. Such a common hardware architecture is denoted by the term *hardware platform*, and consequently the design tasks related to such an approach are grouped under the term *platform-based design* [Keu00].

Once a hardware platform has been fixed, the software functions have to be specified. For the specification of functionality we use a control and dataflow graph based representation [Ele98a], [Ele00] described in detail in Section 2.3.1.

Next, the designer has to decide what part of the functionality should be implemented on which of the selected components (the *mapping* task) and what is the execution order of the resulting functions (the *scheduling* task). An important design task in the context of distributed applications is the *communication synthesis* task, which decides the characteristics of the communication infrastructure and the access constraints to the infrastructure, imposed on functions initiating an inter-node communication.

These design tasks can partially overlap, and they can be assisted by analysis and (semi)automatic synthesis tools. In addition, the design tasks have to be performed such that the timing constraints of hard real-time applications are satisfied, and the implementation costs are minimized.

The analysis takes into account the heterogeneous interconnected nature of the architecture, and is based on an application model which captures both the dataflow and the flow of control. The synthesis techniques described here derive implementations of the system that fulfill the design constraints and reduce the costs. An important part of the system implementation is the synthesis of the communication infrastructure, which has a significant impact on the overall system performance and cost.

The design of real-time systems seldom starts from scratch. Typically, designers start from an already existing system, running certain applications, and the design problem is to implement new functionality on this system. Moreover, after the new functionality has been implemented, the resulting system has to be structured such that additional functionality, later to be added, can easily be accommodated.

Such an approach provides a high degree of flexibility during the design process, and thus, can result in important reductions of design costs. Therefore, the analysis and synthesis methods presented have been considered within such an *incremental design process*.

This book is structured in four parts, and has ten chapters. In the first part we present the application model and the hardware and software architectures considered. The second part presents analysis and synthesis methods for *time-driven* systems, where the activation of processes and transmission of messages happen at predetermined points in time. In the third part we provide an analysis and develop techniques for the synthesis of *event-driven* systems, where the activation of processes is done at the occurrence of significant events. The fourth part combines time-driven and event-driven systems into heterogeneous networks, and presents analysis and synthesis methods for applications distributed across such networks.

This is, briefly, what each chapter is about:

- Part I: Preliminaries
  - Chapter 2 (System-Level Design and Modeling) presents the design methodologies commonly used in embedded systems design, with an emphasis on function/architecture co-design. We introduce the application model considered, a control and dataflow graph based representation [Ele98a], [Ele00] called *conditional process graph*, as well as a model for characterizing existing and future applications within an incremental design process.
  - Chapter 3 (Distributed Hard Real-Time Systems) presents the time-driven and event-driven approaches to the design of real-time systems and introduces the non-preemptive static cyclic scheduling and fixed priority preemptive scheduling policies. We also present the hardware and software architectures considered, including the details of the communication protocols used: the *time triggered protocol* (TTP) [Kop03], which is a time-driven protocol based on a *time-division multiple access* (TDMA) bus access scheme, and the *controller area network protocol* (CAN) [Bos91], an event-driven communication protocol employing a collision avoidance scheme.
- Part II: Time-Driven Systems
  - Chapter 4 (Scheduling and Bus Access Optimization for Time-Driven Systems) considers a non-preemptive static scheduling approach for both processes and messages. In such a context, we discuss the static cyclic scheduling of systems with data and control dependencies. The presented technique is then extended to handle the scheduling of messages over a communication channel using the time-triggered protocol. Several approaches to the syn-

thesis of communication parameters of a TDMA bus are proposed.
  - Chapter 5 (Incremental Mapping for Time-Driven Systems) investigates the mapping and scheduling tasks in the context of an incremental design approach. Such a design process satisfies two main requirements when adding new functionality: the already running applications are disturbed as little as possible, and there is a good chance that, later, new functionality can easily be mapped on the resulted system.

- Part III: Event-Driven Systems
  - Chapter 6 (Schedulability Analysis and Bus Access Optimization for Event-Driven Systems) assumes a preemptive fixed priority scheduling approach for the processes and a non-preemptive static cyclic scheduling approach for the messages, based on the TTP. A schedulability analysis is developed considering four message scheduling approaches for TTP. In addition, we show how, by considering both data and control dependencies when modeling an application, it is possible to reduce the pessimism of the analysis. Several optimization strategies that derive the parameters of the communication protocol are also presented.
  - Chapter 7 (Incremental Mapping for Event-Driven Systems) addresses the same mapping and scheduling problems inside an incremental design process as discussed in Chapter 5, but this time in the context of the architectures and event-driven scheduling policies considered in Chapter 6.

- Part IV: Multi-Cluster Systems
  - Chapter 8 (Schedulability Analysis and Bus Access Optimization for Multi-Cluster Systems) introduces the concept of *multi-cluster systems*, which are heterogeneous networks composed of several networks (called *clusters*),

interconnected via gateways. In this chapter we consider a two-cluster configuration composed of a time-driven cluster and an event-driven cluster. We propose a schedulability analysis technique for such systems, which also produces bounds on the communication buffer sizes required by an application to fulfill its timing constraints. Optimization strategies are developed, aiming at synthesizing a communication infrastructure that would guarantee the timing constraints of the application at the same time with minimizing the system implementation costs.

— Chapter 9 (Partitioning and Mapping for Multi-Cluster Systems) addresses design problems which are characteristic to multi-clusters: partitioning of the system functionality into time-triggered and event-triggered domains, and mapping of processes to the heterogeneous nodes of a cluster. We present several heuristics for solving these problems, and show that they are able to find schedulable implementations under limited resources, achieving an efficient utilization of the system.

— Chapter 10 (Schedulability-Driven Frame Packing for Multi-Cluster Systems) addresses the issue of packing of messages to frames in the case of TTP and CAN protocols. We have updated the schedulability analysis presented in Chapter 8 to take into account the details related to frames, and we will discuss two optimization heuristics that use the schedulability analysis as a driver towards a frame configuration that leads to a schedulable system.

All the approaches presented in the book have been evaluated using an extensive set of applications generated for experimental purposes. In addition, we also use throughout the book a real-life case study, a vehicle cruise controller, in order to illustrate the relevance of the identified problems and to further evaluate the proposed approaches.

Chapter 2
System-Level Design and Modeling

THE MODELING AND design of embedded systems can be performed at several abstraction levels. Gajsky [Gaj83] identifies the following abstraction levels in the context of CAD tools for VLSI:

- *Circuit level* is the lowest level of abstraction. For example, the hardware at this level is seen as transistors, capacitors, resistors, etc., and differential equations are often used to describe their functionality.
- *Logic level* is next towards higher levels of abstraction. Here, the functionality is represented as boolean logic (hence the name, *logic* level), implemented in hardware using logic gates and flip-flops.
- At the *register-transfer level* the functionality is captured in terms of register-transfer operations on ALUs, registers, multiplexers, etc.
- The highest level of abstraction is the *system level*, where the functionality is described using "system-level specification formalisms" (in the case of VLSI design, these can be description languages like VHDL, Verilog or SystemC) and the archi-

tecture is seen as building blocks consisting of processors, memories, etc., interconnected using buses.

The research presented in this book is dealing with the design issues at the system level of abstraction. Providing methodologies, techniques and tools for system-level design is the only solution to the increasing complexity of embedded systems, and the designer's productivity gap [Sem02]. The system-level design methodology is presented in the next section. In our research, we place a special emphasis on an incremental design process as outlined in Section 2.2.

At system level, we view the architectures as a set of heterogeneous interconnected networks, each network consisting of nodes sharing the same communication channel. Each node has a processor, a memory, a communication controller, and I/Os to sensors and actuators. Also, the functionality is captured as a set of interacting processes, modeled using a process network formalism. The exact representation we use for modeling the functionality at the system level is described in Section 2.3, while the architectures considered are described in more detail in Chapter 3.

2.1 System-Level Design

The aim of a design methodology is to coordinate the design tasks such that the time-to-market is minimized, the design constraints are satisfied, and various parameters are optimized. System-level design is illustrated in Figure 2.1 (adapted from [Ele02]). It emphasizes the design tasks that happen before the hardware and software components are definitively identified. According to the figure, which groups the system-level design tasks and models inside the grey rectangle, system-level design tasks take as input the specification of the system and its requirements, and produce the hardware and software models, which are later synthesized.

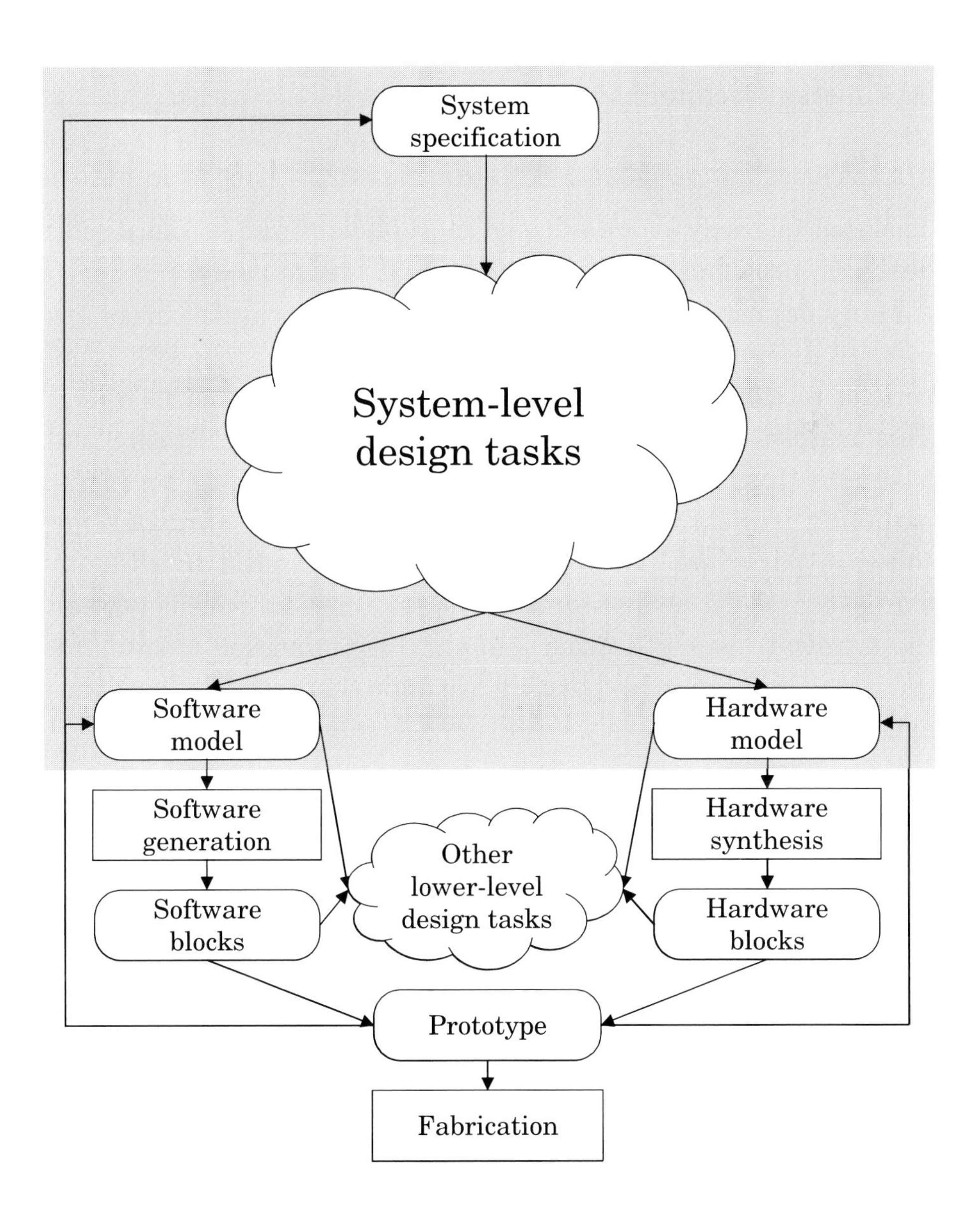

Figure 2.1: System-Level Design

In the next sections we discuss some approaches to system-level design, namely:

1. Traditional design methodology,
2. Hardware/software co-design, and
3. Function/architecture co-design.

2.1.1 TRADITIONAL DESIGN METHODOLOGY

This methodology is not a design methodology per se, but a set of design approaches traditionally used in the industry.

Many organizations, including automotive manufacturers, are used to designing and developing their systems following some version of the *waterfall* model of system development [Wol01]. This means that the design process starts with a specification and, based on this, several system-level design tasks are performed manually, usually in an *ad-hoc* fashion. Then, the hardware and software parts are developed independently, often by different teams located far away from each other. Software code is written, the hardware is synthesized and they are supposed to integrate correctly. Simulation and testing are done separately on hardware and software, respectively, with very few integration tests.

If this design approach was appropriate when used for relatively small systems produced in a well defined production chain, it performs poorly for more complex systems, leading to an increase in the time-to-market. There are several reasons for this. First of all, it is very difficult, just based on the specification, to accurately determine what system architecture is appropriate and how the resources should be used. Also, a separate view on the hardware and software design processes (which are dependent on each other) leads to a poorly designed system, which often has a poor performance because of the incomplete exploration of the trade-offs between the software and hardware domains.

New design methodologies are needed in order to cope with the increasing complexity of current systems.

2.1.2 HARDWARE/SOFTWARE CO-DESIGN

The main idea behind *hardware / software co-design* is to concurrently design (hence the term co-design) and develop both the hardware and the software components of the system. Surveys about hardware/software co-design can be found in [Mic96], [Mic97], [Ern98], [Gaj95], [Sta97], [Wol94], [Wol03].

At the beginning, researchers proposing co-design approaches made quite restrictive assumptions, and the goals were modest. These approaches are not really system level, they actually belong to the "other lower-level design tasks" cloud in Figure 2.1. For example, several researchers have assumed a simple input specification in form of a computer program, and the main goal was to obtain an as high as possible execution performance within a given cost budget (acceleration). The architecture considered consisted of a single processor together with an ASIC used to accelerate parts of the functionality [Cho95a], [Gup95], [Moo97]. In this context, the main problems were to divide the functionality between the ASIC and the CPU (hardware/software partitioning) [Axe96], [Ele97], [Ern93], [Gup93], [Vah94], to automatically generate drivers and other components related to communication (communication synthesis) [Cho92], [Wal94] and to simulate and verify the resulting system (co-simulation and co-verification) [Val95], [Val96].

However, today such restrictive assumptions are no longer valid and the goals are much broader [Bol97], [Dav98], [Dav99], [Dic98], [Lak99], [Ver96]:

- The system specification is now assumed to be inherently heterogeneous and complex. Several languages as well as several models of computation can be found within a specification.
- The architectures are varied, ranging from distributed embedded systems, in the automotive electronics area, to systems on a chip used in telecommunications. The hardware architectures are heterogeneous, consisting of not only

programmable processors and ASICs, but also DSPs, FPGAs, ASIPs, etc.

- The goals include not only acceleration with minimal hardware cost, but also issues related to the reuse of legacy hardware and software subsystems, real-time constraints, quality of service, fault tolerance and dependability, power consumption, flexibility, time-to-market, etc.

2.1.3 FUNCTION/ARCHITECTURE CO-DESIGN

Several researchers [Tab00], [Lav99] have pointed out that most of the hardware/software co-design approaches are not really addressing the design tasks at system-level, but are rather emphasizing on the interaction between the hardware and software entities (the "other lower-level design tasks" in Figure 2.1).

For this reason, a *function/architecture co-design* methodology has been proposed [Kie97], [Lie99], [Tab00], [Lav99], [Bal97], which addresses also the design process before hardware/software partitioning. This move towards even higher abstraction levels has been considered as the key to shortening design time and coping with complexity.

Function/architecture co-design uses a top-down synthesis approach, where trade-offs are evaluated at a high level of abstraction (see Figure 2.2, adapted from [Ele02]). The main characteristic of this methodology is the use, at the same time with the top-down synthesis, of a bottom-up evaluation of design alternatives, without the need to perform a full synthesis of the design. The approach to obtaining accurate evaluations is to use an accurate modeling of the behavior and architecture (the "Mapped and scheduled model" box in Figure 2.2), and to develop analysis techniques that are able to derive estimates and to formally verify properties relative to a certain design alternative (the "Estimation" and "Simulation and verification" boxes). The determined estimates and properties, together with

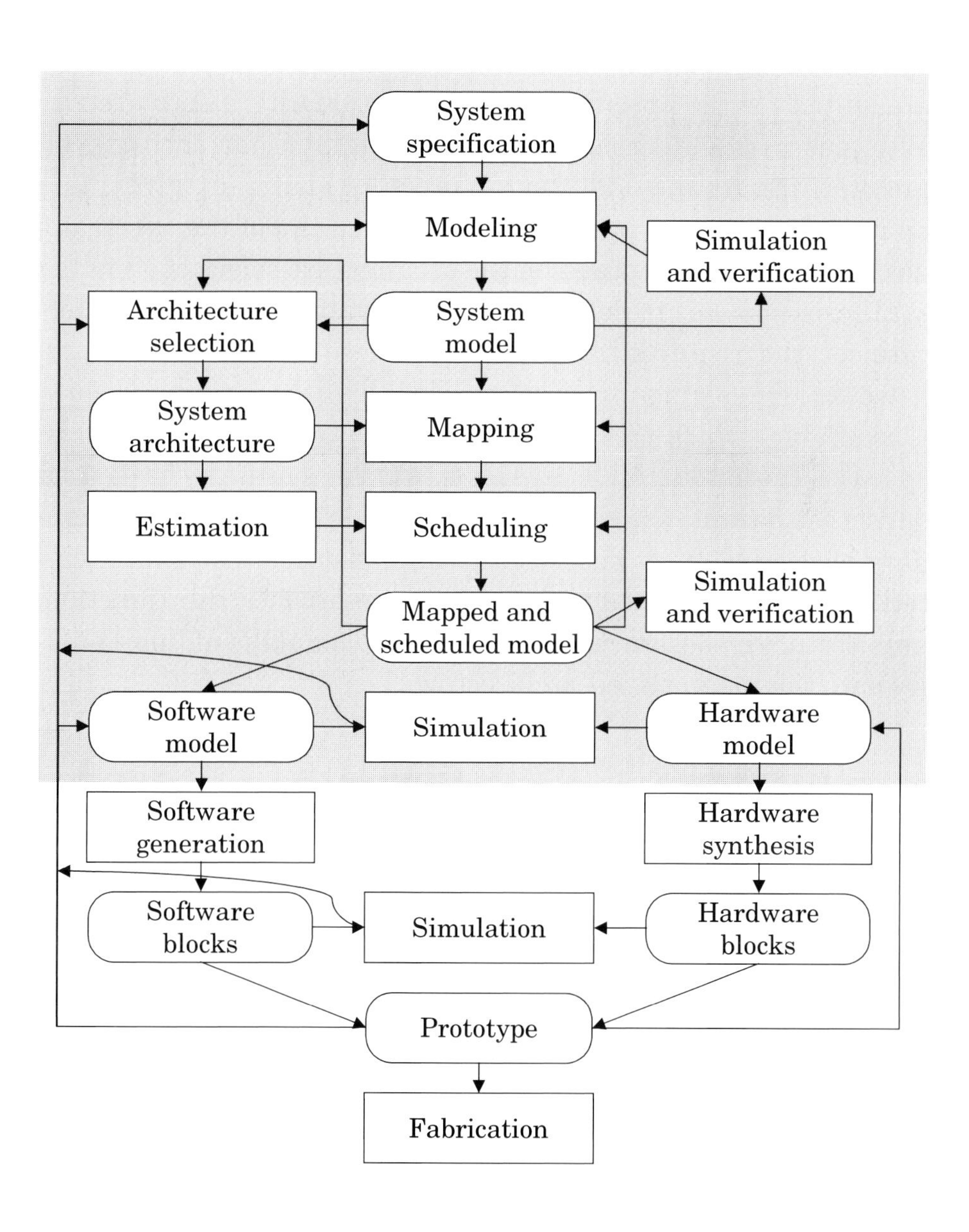

Figure 2.2: Function/Architecture Co-design

user-specified constraints, are then used to drive the synthesis process.

Thus, several architectures are evaluated to determine if they are suited for the specified system functionality. There are two extremes in the degrees of freedom available for choosing an architecture. At one end, the architecture is already given, and no modifications are possible. At the other end of the spectrum, no constraints are imposed on the architecture selection, and the synthesis task has to determine, from scratch, the best architecture for the required functionality. These two situations are, however, not common in practice. Usually, a *hardware platform* is available, which can be *parameterized* (e.g., size of memory, speed of the buses, etc.). In this case, the synthesis task is to derive the parameters of the architecture such that the functionality of the system is successfully implemented. Once an architecture is determined and/or parameterized, the function/architecture co-design continues with the mapping of functionality onto the instantiated architecture.

2.1.4 PLATFORM-BASED DESIGN

As the complexity of the systems continues to increase, the development time lengthens dramatically, and the manufacturing costs become prohibitively high. To cope with this complexity, it is necessary to reuse as much as possible at all levels of the design process, and to work at higher and higher abstraction levels.

One of the most important components of any system design methodology is the definition of a *system platform*. Such a platform consists of a hardware infrastructure together with software components that will be used for several product versions, and will be shared with other product lines, in the hope to reduce costs and the time-to-market.

The authors in [Keu00] have proposed techniques for deriving such a platform for a given family of applications. Their

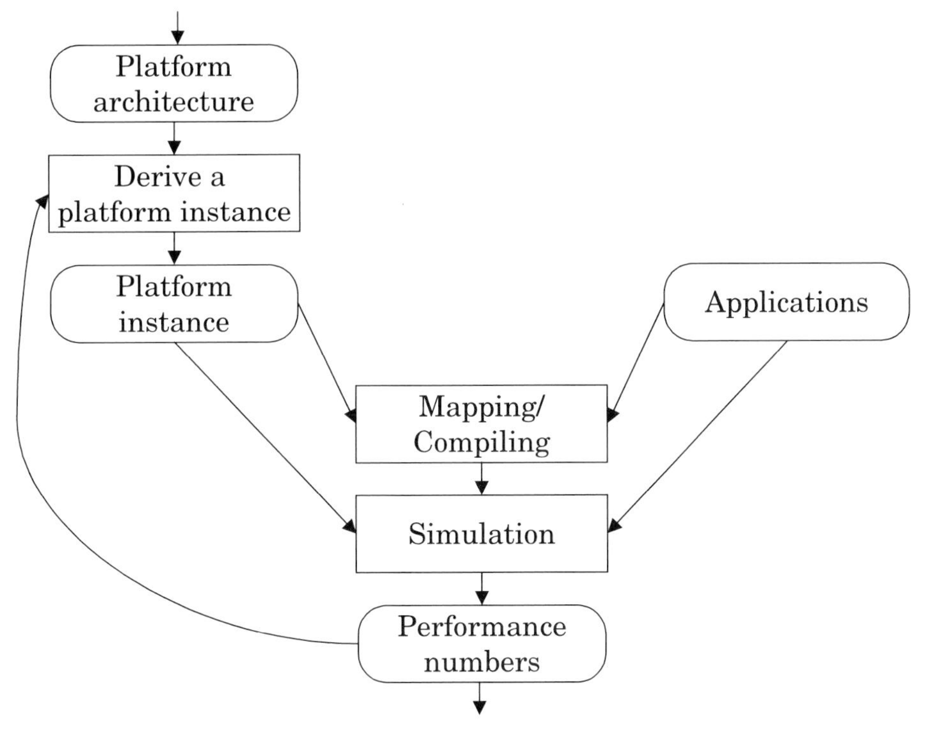

Figure 2.3: Platform-Based Design

approach can be used within any design methodology for determining a system platform that later on can be parameterized and instantiated to a desired hardware architecture.

Considering a given application or family of applications, the hardware platform has to be instantiated, deciding on certain parameters, and lower level details, in order to suit that particular application(s). In Figure 2.3 (adapted from [Kie97]), the search for an architecture instance starts from a certain platform, and a given application. The application is mapped and compiled on an architecture instance, and the performance numbers are derived, typically using simulation. If the designer is satisfied with the performance of the instantiated architecture, the loop ends.

The research presented in this book concentrates on the following system-level design tasks:

1. mapping,
2. scheduling, and
3. communication synthesis.

In addition, we consider these design tasks within an incremental design process, as outlined in the next section.

2.2 Incremental Design Process

A characteristic of the majority of research efforts concerning the design of embedded systems is that the authors concentrate on the design, from scratch, of a new system optimized for a particular application. For many application areas, however, such a situation is extremely uncommon and only rarely appears in design practice. It is much more likely that one has to start from an already existing system running a certain application and the design problem is to implement new functionality (including also upgrades to the existing one) on this system. In such a context it is very important to operate no, or as few as possible, modifications to the already running application. The main reason for this is to avoid unnecessarily large design and testing times. Performing modifications on the (potentially large) existing application increases design time and, even more, testing time (instead of only testing the newly implemented functionality, the old application, or at least a part of it, has also to be retested).

However, this is not the only aspect to be considered. Such an incremental design process, in which a design is periodically upgraded with new features, is going through several iterations. Therefore, after new functionality has been introduced, the resulting system has to be implemented such that additional functionality, later to be mapped, can easily be accommodated.

We illustrate such an incremental design process in Figure 2.4. The product is implemented as a three processor system and its version N–1 consists of the set ψ of two applications (the processes belonging to these applications are represented as white and black disks, respectively). At the current moment, application $\Gamma_{current}$ is to be added to the system, resulting in version N of the product. However, a new version, N+1, is very likely to follow and this fact is to be considered during implementation of $\Gamma_{current}$[1].

If it is not possible to map and schedule $\Gamma_{current}$ without modifying the already running applications, we have to change the scheduling and mapping of some applications in ψ However, even with serious modifications performed on ψ, it is still possible that certain constraints are not satisfied. In this case the hardware architecture has to be changed by, for example, adding a new processor, and the mapping and scheduling procedure for $\Gamma_{current}$ has to be restarted. In this book we do not elaborate on the aspect of adding new resources to the architecture, but will concentrate on the mapping and scheduling aspects. Therefore, we will consider that a possible mapping and scheduling of $\Gamma_{current}$, which satisfies the imposed constraints can be found (with minimizing the modification of the already running applications), and this solution has to be further improved in order to facilitate the implementation of future applications.

1. The design process outlined here also applies when $\Gamma_{current}$ is a new version of an application $\Gamma_{old} \in \psi$ In this case, all the processes and communications belonging to Γ_{old} are eliminated from the running system ψ before starting the mapping and scheduling of $\Gamma_{current}$.

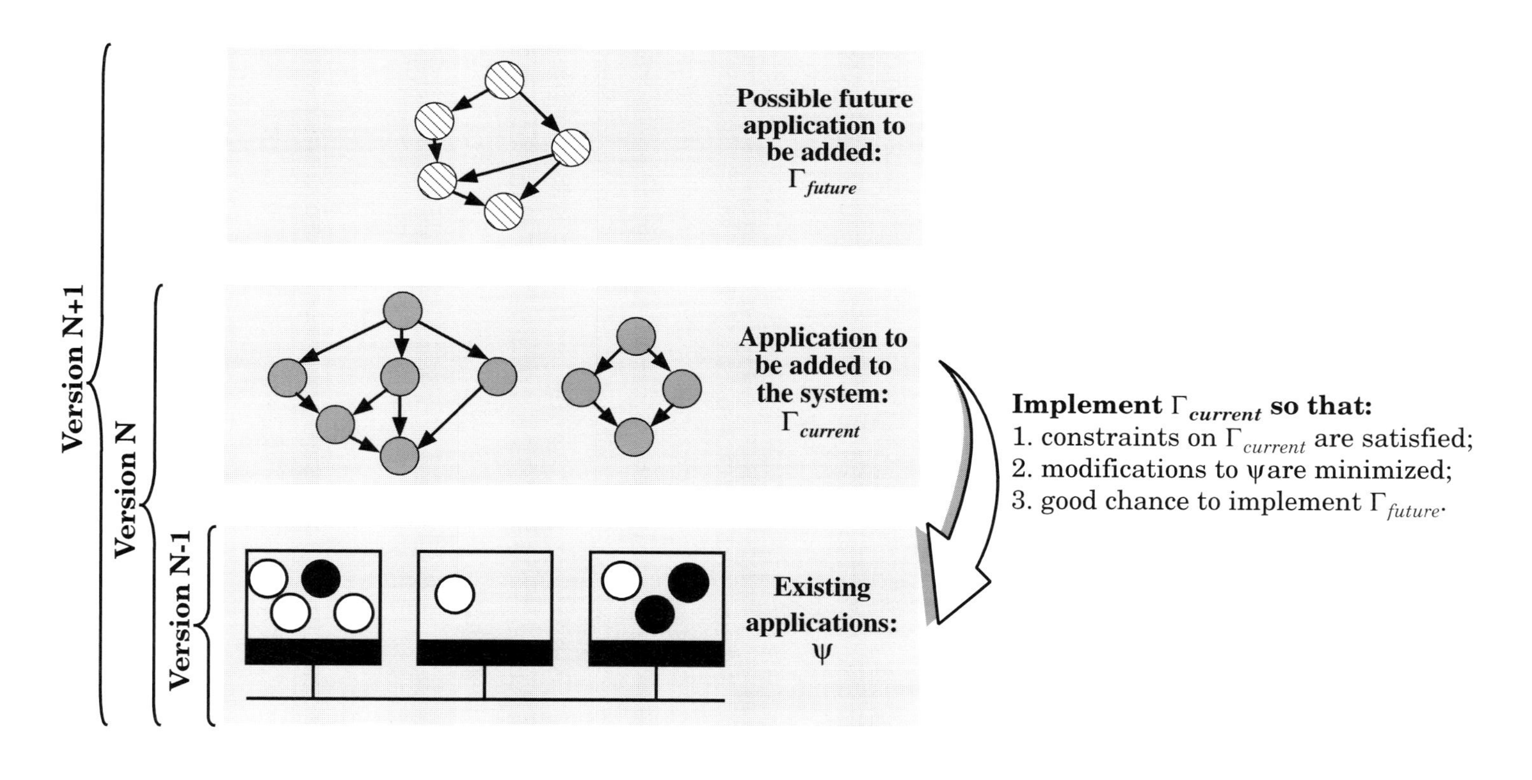

Figure 2.4: Incremental Design Process

2.3 Application Modeling

The functionality of the host system, into which the electronic system is embedded, is normally described using a formalism from that particular domain of application. For example, if the host system is a vehicle, then its functionality is described in terms of control algorithms using differential equations, which are modeling the behavior of the vehicle and its environment. At the level of the embedded system which controls the host system, viewed as the system level for us, the functionality is typically described as a set of functions, accepting certain inputs and producing some output values.

There is a lot of research in the area of system modeling and specification, and an impressive number of representations have been proposed. An overview, classification and comparison of different design representations and modeling approaches is given in [Edw97], [Edw00], [Lav99].

The scheduling and mapping design tasks deal with sets of interacting processes. A *process* is a sequence of computations (corresponding to several building blocks in a programming language) which starts when all its inputs are available. When it finishes executing, the process produces its output values. Researchers have used, for example, *dataflow process networks* (also called *task graphs*, or *process graphs*) [Lee95] to describe interacting processes, and have represented them using directed acyclic graphs, where a node is a process and the directed arcs are dependencies between processes.

One drawback of dataflow process graphs is that they are not suitable to capture the control aspects of an application. For example, it can happen that the execution of some processes can also depend on conditions computed by previously executed processes. By explicitly capturing the control flow in the model, a more fine-tuned modeling and a tighter (less pessimistic) assignment of execution times to processes is possible, compared to traditional data-flow based approaches. Several researchers have proposed extensions to the dataflow process graph model in order to capture these control dependencies [Ele98a], [Thi99], [Kla01].

In this book we use the *conditional process graph* (CPG) [Ele98a], [Ele00] as an abstract model for representing the behavior of the application, as it not only captures both dataflow and the flow of control, but is also suitable for handling timing aspects.

2.3.1 CONDITIONAL PROCESS GRAPH

We model an application Γ as a set of conditional process graphs $\mathcal{G}_i \in \Gamma$. A *conditional process graph* is an abstract representation consisting of a directed, acyclic, polar graph $\mathcal{G}(V, E_S, E_C)$. Each node $P_i \in V$ represents one process. E_S and E_C are the sets of simple and conditional edges, respectively. $E_S \cap E_C = \emptyset$ and $E_S \cup E_C = E$, where E is the set of all edges. An edge $e_{ij} \in E$ from P_i to P_j indicates that the output of P_i is the input of P_j.

The graph is polar, which means that there are two nodes, called *source* and *sink*, that conventionally represent the first and last process. These nodes are introduced as dummy processes, with zero execution time and no resources assigned, so that all other nodes in the graph are successors of the source and predecessors of the sink respectively.

A mapped conditional process graph, $G(V^*, E_S^*, E_C^*, M)$, is generated from a conditional process graph $\mathcal{G}(V, E_S, E_C)$ by inserting additional processes (communication processes) on certain edges and by mapping each process to a given processing element. The mapping of processes $P_i \in V^*$ to processors and buses is given by a function M: $V^* \rightarrow PE$, where $PE = \{pe_1, pe_2, ..., pe_{N_{pe}}\}$ is the set of processing elements. $PE = PP \cup B$, where PP is the set of programmable processors and B is the set of allocated buses. For every process P_i, $M(P_i)$ is the processing element to which P_i is assigned for execution.

In the process graph depicted in Figure 2.5, P_0 and P_{15} are the source and sink nodes, respectively. The nodes denoted $P_1, P_2, ..., P_{14}$ are "ordinary" processes specified by the designer. They are assigned to one of the three programmable processors, as indicated by the shading in the figure. The rest of the nodes are so called *communication processes* and they are represented in Figure 2.5 as solid circles. They are introduced during the generation of the system representation for each connection which links processes mapped to different processors, and model interprocessor communication. All communications in Figure 2.5 are performed on one bus.

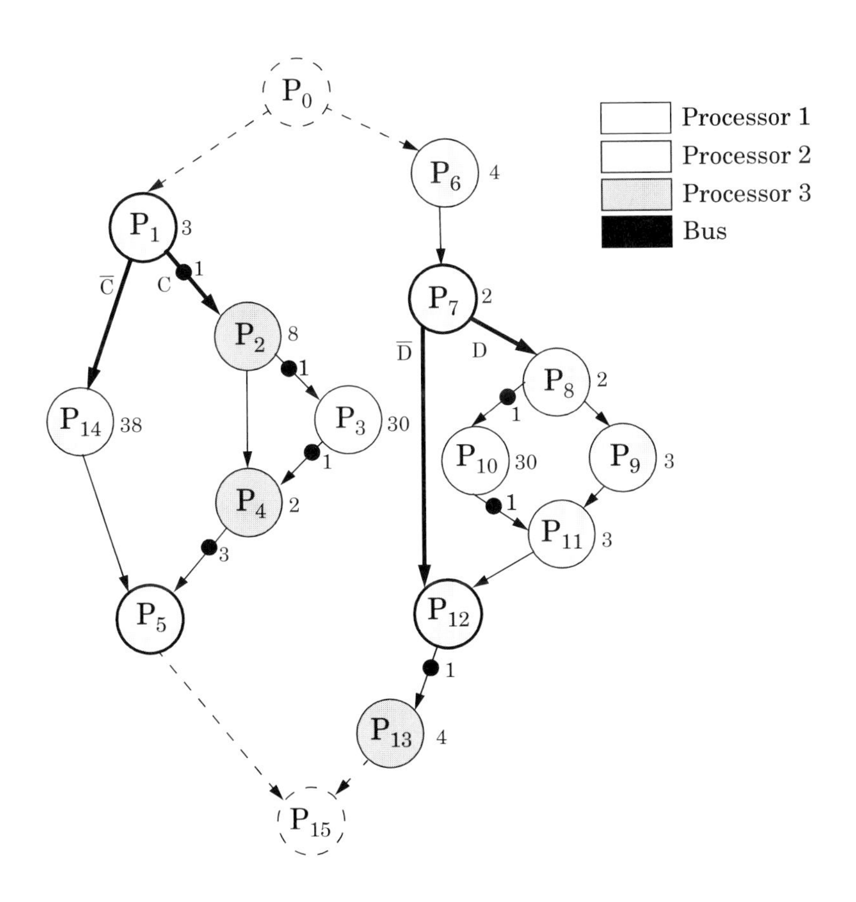

Figure 2.5: A Conditional Process Graph Example

An edge $e_{ij} \in E_C$ is a *conditional edge* (represented with thick lines in Figure 2.5) and has an associated condition value. In Figure 2.5 processes P_1 and P_7 have conditional edges at their output.

We call a node with conditional edges at its output a *disjunction node* (and the corresponding process a *disjunction process*). A disjunction process has one associated condition, the value of which it computes. Alternative paths starting from a disjunction node, which correspond to complementary values of the condition, are disjoint and they meet in a so called *conjunction node* (with the corresponding process called *conjunction process*)[1]. In Figure 2.5, circles representing conjunction and disjunction nodes are depicted with thick borders. The alternative paths starting from disjunction node P_1, which computes condition C, meet in conjunction node P_5. We assume that conditions are independent and alternatives starting from different processes cannot depend on the same condition.

Execution Semantic

The conditional process graph has the following execution semantic:

- A process, that is not a conjunction process, can be activated only after all its inputs have arrived.
- A conjunction process can be activated after messages coming on one of the alternative paths have arrived.
- All processes issue their outputs when they terminate.
- A boolean expression X_{P_i}, called a *guard*, can be associated to each node P_i in the graph. It represents the necessary conditions for the respective process to be activated. X_{P_i} is not only necessary but also sufficient for process P_i to be activated

1. If no process is specified on an alternative path, it is modeled by a conditional edge from the disjunction to the corresponding conjunction node (a communication process may be inserted on this edge at mapping).

during a given system execution. Thus, two nodes P_i and P_j, where P_j is not a conjunction node, are connected by an edge e_{ij} only if $X_{P_j} \rightarrow X_{P_i}$ (which means that X_{P_i} is true whenever X_{P_j} is true). This avoids specifications in which a process is blocked even if its guard is true, because it waits for a message from a process which will not be activated. If P_j is a conjunction node, predecessor nodes P_i can be situated on alternative paths corresponding to a condition.

- Transmission on conditional edges takes place only if the associated condition value is *true* and not, like on simple edges, for each activation of the input process P_i.
- We consider two possible execution environments for processes: *non pre-emptive* and *pre-emptive*:
  - In a non-preemptive environment a process cannot be interrupted during its execution.
  - In a preemptive execution environment a higher priority processes can interrupt the execution of lower priority processes. Also, under certain circumstances, a lower priority process can block a higher priority process (e.g., it is in its critical section), and we consider that the blocking time is computed using the analysis from [Sha90] for the priority ceiling protocol.

The above execution semantic is that of a so called single rate system. It assumes that a node is executed at most once for each activation of the system. If communicating processes with different periods have to be handled (in which case we consider that each process P_i has an associated period T_i), this can be solved by generating several instances of the processes and building a CPG which corresponds to a set of processes as they occur within a time period that is equal to the least common multiple of the periods of the involved processes (see Figure 5.5 on page 111).

Throughout the book we will assume, without loss of generality, that all processes and messages belonging to a process graph G_i have the same period $T_i = T_{G_i}$ which is the period of the process graph.

Specifying Timing Information

Let $\mathcal{N}_{P_i}$ be the set of processing elements to which P_i can potentially be mapped. For each processing element $pe_j \in \mathcal{N}_{P_i}$, we know the worst-case execution time $C_i^{pe_j}$ of process P_i, when executed on pe_j. When the mapping of a process P_i is clear from the context, we use the term C_i to denote its worst-case execution time. In addition, each process P_i is characterized by a period T_i and a priority $priority_{P_i}$[1].

The communication processes (messages), modeling inter-processor communication, have an associated execution time $C_{i,j}$ (where P_i is the sender and P_j the receiver process) equal to the corresponding transmission time. For each message m we know its size S_m. A message is sent once in every n_m invocations of the sending process, with a period $T_m = n_m T_i$ inherited from the sender process P_i. In the case of an event-driven communication protocol (e.g., CAN) messages also have an associated priority, $priority_m$.

As mentioned, we consider execution times of processes, as well as the communication times, to be given. In Figure 2.5 they are depicted to the right of each node. In the case of hard real-time systems this will, typically, be worst case execution times and their estimation has been extensively discussed in the literature [Eng99], [Ern97], [Li95], [Lun99], [Mal97], [Wol02].

If we consider the activation time of the source process as a reference, the activation time of the sink process is the delay of the system at a certain execution. This delay has to be, in the worst case, smaller than a certain imposed deadline D_{G_i} on the process graph G_i. Throughout the book we assume that the deadline can be larger than the period.

Deadlines can also be placed locally on processes. Release times of some processes as well as multiple deadlines can be eas-

1. In the case of a static cyclic scheduling environment no priority has to be attached to the process.

ily modeled by inserting dummy nodes between certain processes and the source or the sink node respectively. These dummy nodes represent processes with a certain execution time but which are not allocated to any processing element.

2.3.2 INCREMENTAL DESIGN: MODELING THE ALREADY IMPLEMENTED APPLICATIONS

Let us consider an incremental design process as outlined in Section 2.2 (Figure 2.4). If the initial attempt to schedule and map $\Gamma_{current}$ does not succeed, we have to modify the schedule and, possibly, the mapping of some already running applications, belonging to ψ in the hope to find a valid solution for $\Gamma_{current}$.

The goal is to find that minimal modification to the existing system which leads to a correct implementation of $\Gamma_{current}$. In our context, such a minimal modification means remapping and/or rescheduling a subset Ω of the old applications, $\Omega \subseteq \psi$, so that the total cost of re-implementing Ω is minimized.

Remapping and/or rescheduling a certain application $\Gamma_i \in \psi$ can trigger the need to also perform modifications to one or several other applications because of, for example, the dependencies between processes belonging to these applications. In order to capture such dependencies between the applications in ψ as well as their modification costs, we have introduced a representation called the *application graph*.

We represent a set of applications as a directed acyclic graph $\mathcal{A}(\mathcal{V}, \mathcal{E})$, where each node $\Gamma_i \in \mathcal{V}$ represents an application. An edge $e_{ij} \in \mathcal{E}$ from Γ_i to Γ_j indicates that any modification to Γ_i would trigger the need to also remap and/or reschedule Γ_j, because of certain interactions between the applications[1]. Each application in the graph has an associated attribute specifying if

1. If a set of applications have a circular dependence, such that the modification of any one implies the remapping of all the others in that set, the set will be represented as a single node in the graph.

that particular application is allowed to be modified or not (in the last case, it is called “frozen”). To those nodes $\Gamma_i \in \mathcal{V}$ representing modifiable applications, the designer has associated a cost R_{Γ_i} of re-implementing Γ_i. Given a subset of applications $\Omega \subseteq \psi$ the total cost of modifying the applications in Ω is:

$$R(\Omega) = \sum_{\Gamma_i \in \Omega} R_{\Gamma_i} \quad (2.1)$$

Modifications of an already running application can only be performed if the process graphs corresponding to that application, as well as the related deadlines (which have to be satisfied also after re-mapping and re-scheduling), are available. However, this is not always the case, and in such situations that particular application has to be considered frozen.

Example 2.1: In Figure 2.6 we present the application graph corresponding to a set of ten applications. Applications Γ_6, Γ_8, Γ_9 and Γ_{10}, depicted in black, are frozen: no modifications are possible to them. The rest of the applications have the modification cost R_{Γ_i} depicted on their left. For example, Γ_7 can be remapped with a cost of 20. If Γ_4 is to be re-implemented, this also requires the modification of Γ_7, with a total cost of 90. In the case of Γ_5, although not frozen, no remap-

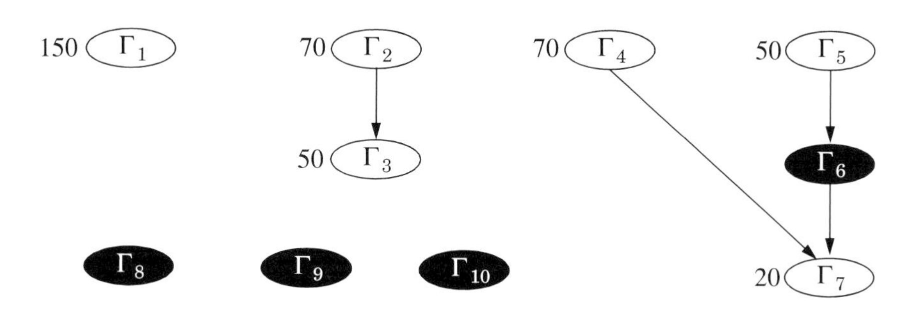

Figure 2.6: Modeling Already Implemented Applications

ping or rescheduling is possible as it would trigger the need to modify Γ_6, which is frozen.

■

As mentioned before, to each application $\Gamma_i \in \mathcal{V}$ the designer has associated a cost R_{Γ_i} of re-implementing Γ_i. Such a cost can typically be expressed in man-hours needed to perform retesting of Γ_i and other tasks connected to the remapping and rescheduling of the application.

If an application is remapped or rescheduled, it has to be validated again. Such a validation phase is very time consuming. In the automotive industry, for example, the time-to-market in the case of the powertrain unit is 24 months. Out of these, 5 months, representing more than 20%, are dedicated to validation. In the case of the telematic unit, the time to market is less than one year, while the validation time is two months [San03]. However, if an application is not modified during implementation of new functionality, only a small part of the validation tasks have to be re-performed (e.g., integration testing), thus reducing significantly the time-to-market, at no additional hardware or development cost.

How to concretely perform the estimation of the modification cost related to an application is beyond the topics of this book. Several approaches to cost estimation for different phases of the software life-cycle have been elaborated and are available in the literature [Deb97], [Rag02]. One of the most influential software cost models is the Constructive Cost Model (COCOMO) [Boe00]. COCOMO is at the core of tools such as REVIC [REV94] and its newer version SoftEST [Sof97], which can produce cost estimations not only for the total development but also for testing, integration, or modification related retesting of embedded software. The results of such estimations can be used by the designer as the cost metrics assigned to the nodes of an application graph.

In general, it can be the case that several alternative costs are associated to a certain application, depending on the particular modification performed. Thus, for example, we can have a certain cost if processes are only rescheduled, and another one if they are also remapped on an alternative node. For different modification alternatives considered during design space exploration, the corresponding modification cost has to be selected. In order to keep the discussion reasonably simple, throughout the book we will consider the case with one single modification cost associated to an application. However, the generalization for several alternative modification costs is straightforward.

2.3.3 CRUISE CONTROLLER EXAMPLE

A typical safety critical application with hard real-time constraints, to be implemented on a distributed architecture, is a vehicle cruise controller (CC).

The CC described in this specification delivers the following functionality:

- It maintains a constant speed for speeds over 35 km/h and under 200 km/h.
- It offers an interface (buttons) to increase or decrease the reference speed.
- It is able to resume its operation at the previous reference speed.
- The CC operation is suspended when the driver presses the brake pedal.

It is assumed that the CC will operate in a distributed environment consisting of several interconnected nodes. There are five nodes which functionally interact with the CC system: the Antilock Braking System (ABS), the Transmission Control Module (TCM), the Engine Control Module (ECM), the Electronic Throttle Module (ETM), and the Central Electronic Module (CEM).

We have considered two hardware architectures for the implementation of the cruise controller, presented in Figure 2.7. In Figure 2.7a, all nodes are connected to a TTP bus, while in Figure 2.7b, we have a two cluster system. In Figure 2.7b, the

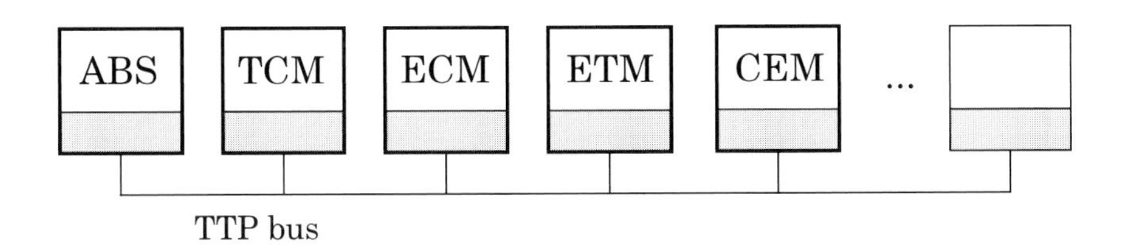

a) Hardware architecture: five nodes connected by a TTP bus

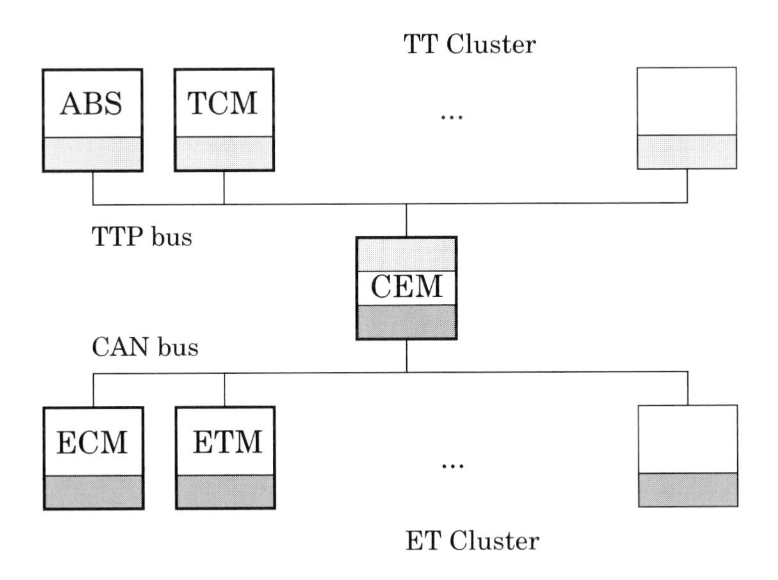

b) Hardware architecture: a two cluster system

Figure 2.7: Two Hardware Architectures for the Cruise Controller

ABS and TCM nodes are part of the time-triggered cluster, that uses the TTP as the communication protocol, and the ECM and ETM nodes are on the event-triggered cluster, which uses CAN. The CEM node is the gateway, connected to both networks.

We have modeled the specification of the CC system using a conditional process graph that consists of 32 processes, and includes two conditions. The first condition, calculated by the source node, decides if the CC is in operation or not (ON or OFF), while the second condition is used to decide, in the case the CC is operational, if the car should speedup or break when trying to reach the reference speed.

The model is presented in Figure 2.8 without assuming any mapping. However, when discussing the scheduling tasks addressed in this book, the mapping is considered as already given. For those cases, we will use the mapping depicted in Figure 2.9. In addition to the nodes representing processes, in Figure 2.9 we have also introduced nodes representing the communication between processes mapped on different processors, depicted with black dots (see Section 2.3.1). The worst-case execution times, considering the mapping in the figure, are depicted to the right of each process. The message sizes are depicted to the left of each message.

The cruise controller example presented in this section will be used in the following chapters for evaluating different approaches to the discussed problems.

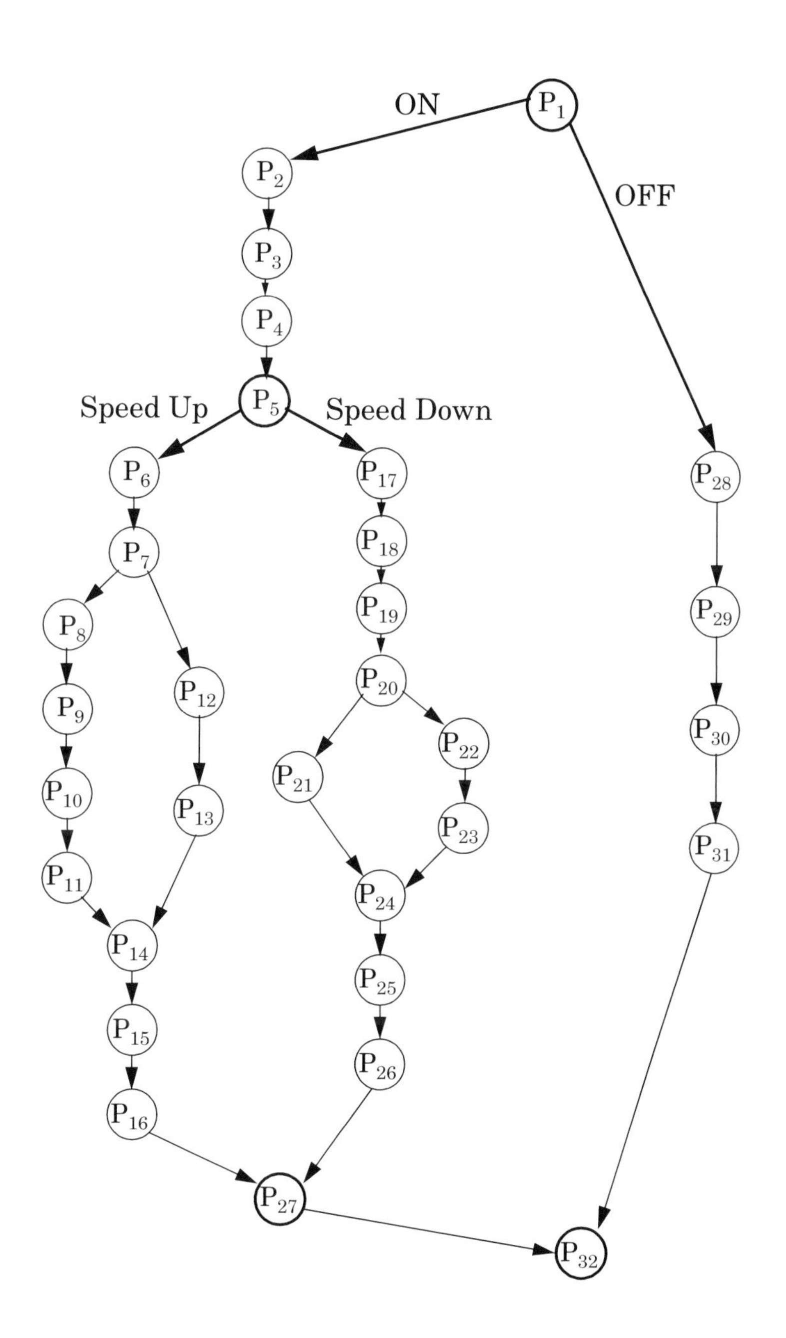

Figure 2.8: The Cruise Controller Model

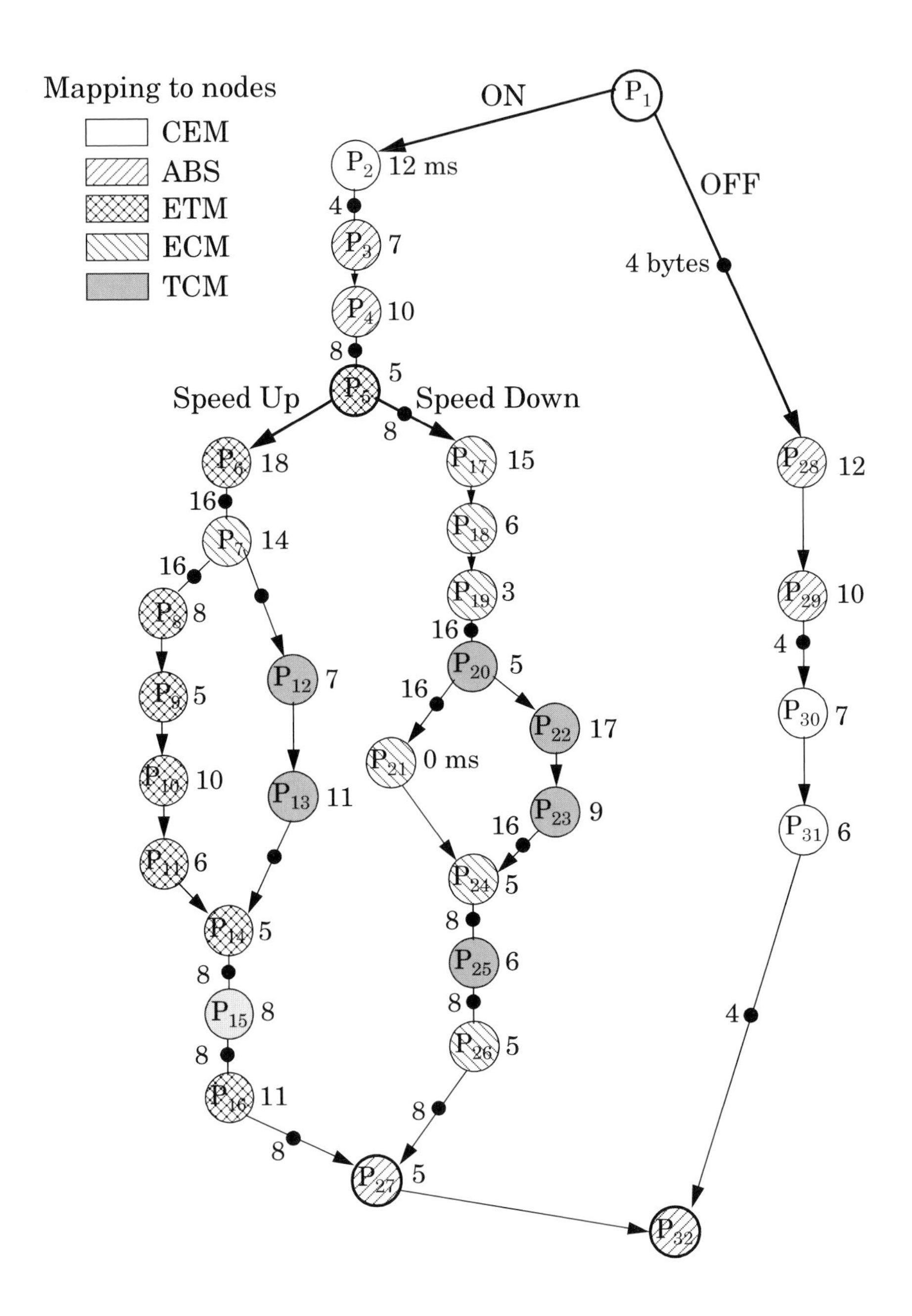

Figure 2.9: The Mapping of the Cruise Controller Model

Chapter 3
Distributed Hard Real-Time Systems

DEPENDING ON THE particular application, real-time systems can be implemented as uniprocessor, multiprocessor, or distributed systems. These systems can be hard or soft, event-driven or time-driven, fault-tolerant, autonomous, etc. A good classification of real-time systems is given in [Kop97a].

Our discussion in this book concentrates on safety-critical hard real-time applications implemented on distributed platforms, where communication has an important impact on the global functionality.

This chapter describes the hardware and software architectures we consider for the implementation of a distributed real-time system. The general hardware platform is introduced in Section 3.2. We particularize this platform for time-driven systems in Section 3.3, present event-driven systems in Section 3.4, and show how the two can be combined into multi-cluster systems in the last section.

3.1 Time-Triggered vs. Event-Triggered

According to [Kop97a] a *trigger* is "*an event that causes the start of some action, e.g., the execution of a task or the transmission of a message.*" Two different approaches to the design of real-time systems can be identified, based on the triggering mechanisms for the processing and communication:

- *Time-Triggered* (TT)
  In the time-triggered approach activities are initiated at predetermined points in time. In a distributed time-triggered system it is assumed that the clocks of all nodes are synchronized to provide a global notion of time. Time-triggered systems are typically implemented using *non-preemptive static cyclic scheduling*, where the process activation or message communication is done based on a schedule table built off-line.
- *Event-Triggered* (ET)
  In the event-triggered approach activities happen when a significant change of state occurs. Event-triggered systems are typically implemented using *preemptive priority-based scheduling*, where, as response to an event, the appropriate process is invoked to service it.

There has been a long debate in the real-time and embedded systems communities concerning the advantages of each approach [Aud93], [Kop97a], [Xu93]. Several aspects have been considered in favour of one or the other approach, such as flexibility, predictability, jitter control, processor utilization, and testability.

An interesting comparison of the ET and TT approaches, from a more industrial, in particular automotive, perspective, can be found in [Lön99]. The conclusion there is that one has to choose the right approach, depending on the particularities of the application.

The analysis and synthesis techniques presented in this book are applied to the following three types of systems. The hardware architectures for these three types of systems are particularizations of the generic hardware platform presented in the next section.

1. *Time-Driven Systems*
   In time-driven systems processes are time-triggered. The details of the hardware and software architectures for time-driven systems are presented in Section 3.3. The mapping, scheduling and communication synthesis methods for time-driven applications are presented in Part II of this book.
2. *Event-Driven Systems*
   In this type of systems processes are event-triggered. The hardware and software architectures for event-driven systems are detailed in Section 3.4. Part III presents our analysis and synthesis methods for event-driven systems.
3. *Multi-Cluster Systems*
   The systems presented at points one and two are either TT or ET. However, for certain applications, the two approaches can be used together, some processes being TT and others ET. Moreover, efficient implementation of new, highly sophisticated automotive applications, entails the use of TT process sets together with ET ones implemented on top of complex distributed architectures.

   One approach to the design of such systems is to allow ET and TT processes to share the same processor as well as static (TT) and dynamic (ET) communications to share the same bus. Bus sharing of TT and ET messages is supported by protocols which support both static and dynamic communication [Fle02]. Traian Pop et al. [Pop02b], [Pop02c] have addressed the problem of timing analysis and design optimization for such systems.

   In this book, we consider systems designed as interconnected clusters of processors. Each such cluster can be either

TT or ET. Depending on their particular nature, certain parts of an application can be mapped on processors belonging to an ET cluster or a TT cluster.

The hardware and software architectures for such multi-cluster systems are presented in Section 3.5. The analysis and synthesis methods for multi-cluster systems are outlined in Part IV of the book.

3.2 The Hardware Platform

We consider, in the most general case, a hardware platform composed of several networks, interconnected with each other (see Figure 3.1). Each network has its own communication protocol, and internetwork communication is via a *gateway* which is a node connected to both networks. The architecture can contain several such networks, having different types of topologies.

A network is composed of several different types of hardware components, called *nodes*. Every node consists of a communication controller, a CPU, a RAM, a ROM and an I/O interface to sen-

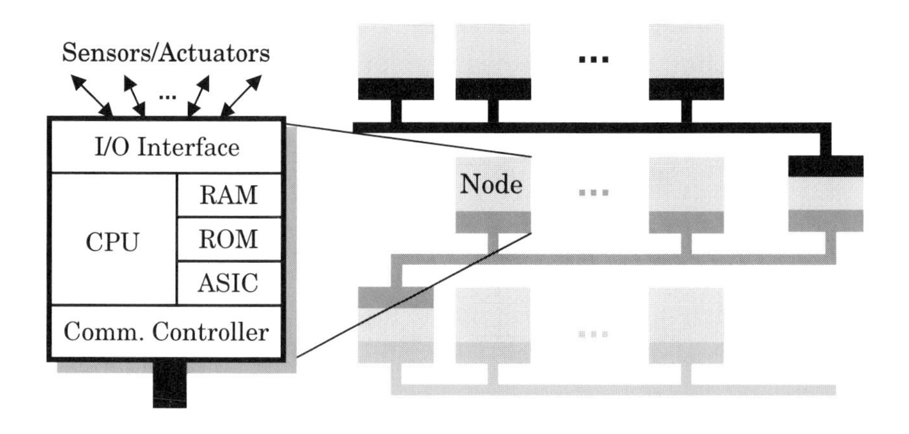

Figure 3.1: Distributed Hard Real-Time Systems

sors and actuators. A node can also have an ASIC in order to accelerate parts of its functionality. The communication controllers implement the protocol services, and run independently of the node's CPU.

The microcontrollers used in a node and the type of network protocol employed are influenced by the nature of the functionality and the imposed real-time, fault-tolerance and power constraints. In the automotive electronics area the functionality is typically divided in two classes, depending on the level of criticalness:

- *Body electronics* refers to the functionality that controls simple devices such as the lights, the mirrors, the windows, the dashboard. The constraints of the body electronic functions are determined by the reaction time of the human operator that is in the range of 100 ms to 200 ms. A typical body electronics system within a vehicle consists of a network of ten to twenty nodes that are interconnected by a low bandwidth communication network like LIN. A node is usually implemented using a single-chip 8 bit micro-controller (e.g., Motorola 68HC05 or Motorola 68HC11) with some hundred bytes of RAM and Kilobytes of ROM, I/O points to connect sensors and to control actuators, and a simple network interface. Moreover, the memory size is growing by more than 25% each year [Kop99], [Chi96].
- *System Electronics* are concerned with the control of vehicle functions that are related to the movement of the vehicle. Examples of system electronics applications are engine control, braking, suspension, vehicle dynamics control. The timing constraints of system electronic functions are in the range of a couple of ms to 20 ms, requiring 16-bit or 32-bit microcontrollers (e.g., Motorola 68332) with about 16 Kilobytes of RAM and 256 Kilobytes of ROM. These microcontrollers have built-in communication controllers (e.g., the 68HC11 and 68HC12 automotive family of microcontrollers

have an on-chip CAN controller), I/O to sensors and actuators, and are interconnected by high bandwidth networks [Kop99], [Chi96].

In order to provide accurate analysis techniques, we need to know the details of the communication protocols used to connect the components of the architecture.

A large number of communication protocols are currently available for embedded systems. However, only a few of them are suitable for safety-critical applications where predictability is mandatory [Rus01]. A survey and comparison of communication protocols for safety-critical embedded systems is available in [Rus01].

The duality between event-triggered and time-triggered approaches discussed in Section 3.1 is also reflected at the level of the communication infrastructure, where communication activities can be triggered either dynamically, in response to an event, or statically, at predetermined moments in time.

Therefore, on one hand, there are protocols that schedule the messages statically based on the progression of time, for example, the SAFEbus [Hoy92] and SPIDER [Min00] protocols for the avionics industry, and the TTCAN [Int02] and Time-Triggered Protocol (TTP) [Kop03] intended for the automotive industry. Out of these, Rushby concludes that TTP "*is unique in being used for both automobile applications, where volume manufacture leads to very low prices, and aircraft, where a mature tradition of design and certification for flight-critical electronics provides strong scrutiny of arguments for safety*" [Rus01].

On the other hand, there are several communication protocols where message scheduling is performed dynamically, such as Byteflight [Ber00] introduced by BMW for automotive applications, Controller Area Network (CAN) [Bos91] used in a large number of application areas including automotive electronics, LonWorks [Ech03] and Profibus [Pro03] for real-time systems in general, etc. Out of these, CAN is the most well known and widespread event-driven communication protocol in the area of distributed embedded real-time systems.

However, there is also a hybrid type of communication protocols, like the FlexRay protocol [Fle02], that allows the sharing of the bus by event-driven and time-driven messages.

Throughout this book we will use the TTP and CAN as representatives for time-driven and event-driven protocols, respectively. A detailed comparison of TTP and CAN is provided in [Kop01].

3.2.1 THE TIME-TRIGGERED PROTOCOL

The time-triggered protocol [Kop03] was designed for distributed real-time applications that require predictability and reliability (e.g., drive-by-wire [XbW98]). It integrates all the services necessary for fault-tolerant real-time systems. TTP services of importance to our problems are: message transport with acknowledgment and predictable low latency, clock synchronization within the microsecond range and rapid mode changes.

The communication channel is a broadcast channel, so a message sent by a node is received by all the other nodes. The bus access scheme is time-division multiple-access (TDMA) (Figure 3.2). Each node N_i can transmit only during a predetermined time interval, the so called TDMA slot S_i. In such a slot, a node can send several messages packaged in a frame. A sequence of slots corresponding to all the nodes in the architecture is called a TDMA round. A node can have only one slot in a TDMA round. Several TDMA rounds can be combined together in a cycle that is repeated periodically. The sequence and length of the slots are the same for all the TDMA rounds. However, the length and contents of the frames may differ.

Every node has a TTP controller that implements the protocol services, and runs independently of the node's CPU. Communication with the CPU is performed through a so called message base

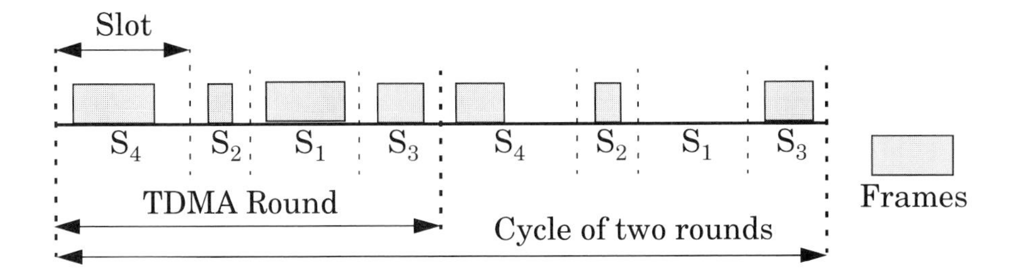

Figure 3.2: TTP Bus Access Scheme

interface (MBI) which is usually implemented as a dual ported RAM (depicted in Figure 3.5).

The TDMA access scheme is imposed by a so called message descriptor list (MEDL) that is located in every TTP controller. The MEDL basically contains the time when a frame has to be sent or received, the address of the frame in the MBI, and the length of the frame. The MEDL serves as a schedule table for the TTP controller which has to know when to send or receive a frame to or from the communication channel.

The TTP controller provides each CPU with a timer interrupt based on a local clock, synchronized with the local clocks of the other nodes. The clock synchronization is done by comparing the a priori known time of arrival of a frame with the observed arrival time. By applying a clock synchronization algorithm, TTP provides a global time-base of known precision, without any overhead on the communication.

There are two types of frames in the TTP. The initialization frames, or I-frames, which are needed for the initialization of a node, and the normal frames, or N-frames, which are the data frames containing, in their data field, the application messages. A TTP data frame (Figure 3.3) consists of the following fields:

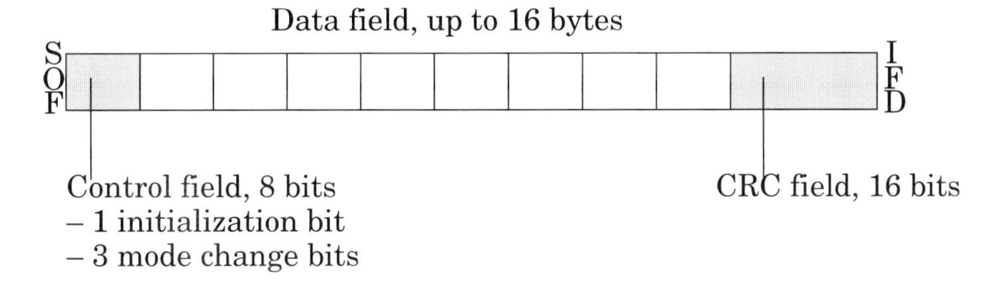

Figure 3.3: TTP Frame Configuration

start of frame bit (SOF), control field, a data field of up to 16 bytes containing one or more messages, and a cyclic redundancy check (CRC) field. Frames are delimited by the inter-frame delimiter (IFD, 3 bits). Note that no identifier bits are necessary, as the TTP controllers know from their MEDL what frame to expect at a given point in time.

In general, the time-triggered protocol efficiency is in the range of 60–80% [Tec02].

> **Example 3.1:** For example, the data efficiency of a frame that carries 8 bytes of application data, i.e., the percentage of transmitted bits which are the actual data bits needed by the application, is 69.5% (64 data bits transmitted in a 92-bit frame, without considering the details of a particular physical layer).
>
> ■

3.2.2 THE CONTROLLER AREA NETWORK PROTOCOL

The controller area network [Bos91] is a priority bus that employs a collision avoidance mechanism, whereby the node that transmits the frame with the highest priority wins the con-

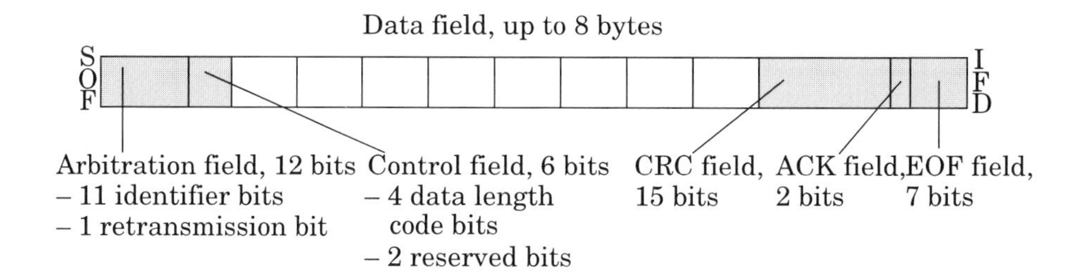

Figure 3.4: CAN 2.0A Data Frame Configuration

tention. Frame priorities are unique and are encoded in the frame identifiers, which are the first bits to be transmitted on the bus.

In the case of CAN 2.0A, there are four frame types: data frame, remote frame, error frame, and overload frame. We are mainly interested in the structure of the data frame, depicted in Figure 3.4. A data frame contains seven fields: SOF, arbitration field that encodes the 11 bit frame identifier, a control field, a data field up to 8 bytes, a CRC field, an acknowledgement (ACK) field, and an end of frame field (EOF).

The typical CAN protocol efficiency is in the range of 25–35% [Tec02].

Example 3.2: For a frame that carries 8 bytes of application data, using the specification of a data frame presented in Figure 3.4, we will have an efficiency of 47.4% [Nol01].

■

3.3 Time-Driven Systems

The first type of systems considered in the book are time-driven systems, in which processes are activated according to a time-triggered policy. Typically, in a time-driven system, messages are transmitted using a time-driven communication protocol such as the TTP, while the scheduling of processes is performed using static cyclic scheduling.

The hardware architecture consists of one single network, composed of a set of nodes interconnected using the TTP. The main component of the software architecture is a real-time kernel that runs on top of each node.

The kernel running as part of the software architecture on each node has a schedule table. This schedule table contains all the information needed to take decisions on activation of processes and transmission of messages, on that particular node.

In order to run a predictable hard real-time application, the overhead of the kernel and the worst case administrative overhead (WCAO) of every system call has to be determined. Having a time-triggered system, all the activity is derived from the progression of time which means that there are no other interrupts except for the timer interrupt.

Several activities, like polling of the I/O or diagnostics, take place directly in the timer interrupt routine. The overhead due to this routine is expressed as the utilization factor U_t. U_t represents a fraction of the CPU power utilized by the timer interrupt routine, and has an influence on the execution times of the processes.

We also have to take into account the overheads for process activation and message passing. For process activation we consider an overhead δ_{PA}. The message passing mechanism is illustrated in Figure 3.5, where we have three processes, P_1, P_2 and P_3. P_1 and P_2 are mapped to node N_1 that transmits in slot S_1, and P_3 is mapped to node N_2 that transmits in slot S_2. Message m_1 is transmitted between P_1 and P_2, which are on the same node, while message m_2 is transmitted from P_1 to P_3 between the two nodes. We consider that each process has its own memory locations for the messages it sends or receives and that the

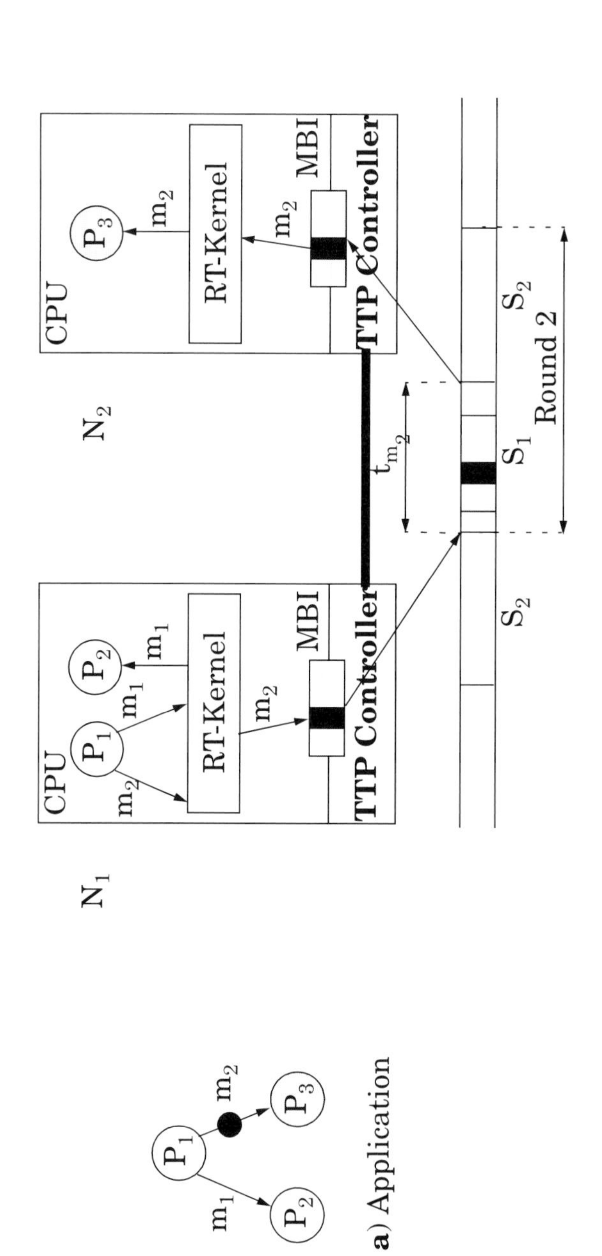

Figure 3.5: A Message Passing Example for Time-Driven Systems

addresses of the memory locations are known to the kernel through the schedule table.

P_1 is activated according to the schedule table, and when it finishes it calls the send kernel function in order to send m_1, and then m_2. Based on the schedule table, the kernel copies m_1 from the corresponding memory location of P_1 to the memory location of P_2. The time needed for this operation represents the WCAO δ_S for sending a message between processes located on the same node. When P_2 will be activated, it will find the message in the right location. According to our scheduling policy, whenever a receiving process needs a message, the message is already placed in the corresponding memory location. Thus, there is no overhead on the receiving side, for messages exchanged on the same node.

Message m_2 has to be sent from node N_1 to node N_2. At a certain time, known from the schedule table, the kernel transfers m_2 to the TTP controller by packing m_2 into a frame in the MBI. The WCAO of this function is δ_{KS}. Later on, the TTP controller knows from its MEDL when it has to take the frame from the MBI, in order to broadcast it on the bus. In our example, the timing information in the schedule table of the kernel and the MEDL is determined in such a way that the broadcasting of the frame is done in the slot S_1 of *Round 2*. The TTP controller of node N_2 knows from its MEDL that it has to read a frame from slot S_1 of *Round 2* and to transfer it into the MBI. The kernel in node N_2 will read the message m_2 from the MBI, with a corresponding WCAO of δ_{KR}[1]. When P_3 will be activated based on the local schedule table of node N_2, it will already have m_2 in its memory location.

1. The overheads δ_S, δ_{KS} and δ_{KR} depend on the length of the transferred message; in order to simplify the presentation this aspect is not discussed further.

3.4 Event-Driven Systems

The second type of systems considered in the book are the event-driven systems in which processes are managed according to an event-driven policy.

The hardware architecture consists of one single network, composed of a set of nodes interconnected using a communication channel.

The scheduling of processes is performed using fixed-priority preemptive scheduling. A natural mapping of event-driven messages would be on a bus implementing an event-triggered protocol at the data-link layer, such as the CAN bus. Such a solution has been considered in literature, for example in [Tin95].

However, considering preemptive priority based scheduling at the process level, with time triggered static scheduling at the communication level can be the right solution under several circumstances [Lön99]. Moreover, a communication protocol like the time-triggered protocol provides a global time base, and improves fault-tolerance and predictability.

Therefore, in Part III of this book we will consider that messages produced by event-triggered processes are transmitted using the time-triggered communication protocol, and we have developed four message passing policies for transmitting event-triggered messages over a time-triggered bus (see Section 6.4). However, for the event-triggered clusters of a multi-cluster system addressed in Part IV, we will consider that the communications are performed using an event-triggered protocol such as the CAN protocol.

As the main component of the software architecture, we have a real-time kernel running on the CPU of each node, which has a scheduler as one of its main components. This scheduler decides on activation of processes, based on their priorities.

As in the previous section, the overhead of the kernel and the worst case administrative overhead (WCAO) of every system call have to be determined. Our schedulability analysis takes into account these overheads, and also the overheads due to the message passing.

The message passing mechanism is illustrated in Figure 3.6, where we have three processes, P_1, P_2, and P_3. As in the example illustrated in Figure 3.5, P_1 and P_2 are mapped to node N_1 that transmits in slot S_1, and P_3 is mapped to node N_2 that transmits in slot S_2. Message m_1 is transmitted between P_1 and P_2 that are on the same node, while message m_2 is transmitted from P_1 to P_3 between the two nodes.

Messages between processes located on the same processor are passed through shared protected objects. The overhead for their communication is accounted for by the blocking factor, using the analysis from [Sha90] for the priority ceiling protocol.

Message m_2 has to be sent from node N_1 to node N_2. Hence, after m_2 is produced by P_1, it will be placed into an outgoing message queue, called *Out*. The access to the queue is guarded by a priority-ceiling semaphore. A so called transfer process (denoted with T in Figure 3.6) moves the message from the *Out* queue into the MBI.

How the message queue is organized and how the message transfer process selects the particular messages and assembles them into a frame depend on the particular approach chosen for message scheduling (see Section 6.4). The message transfer process is activated at certain a priori known moments by the scheduler, in order to perform the message transfer. These activation times are stored in a message handling time table (MHTT) available to the real-time kernel in each node. Both the MEDL and the MHTT are generated off-line as result of the schedulability analysis and optimization which will be discussed later. The MEDL imposes the times when the TTP controller of a certain node has to move frames from the MBI to the communication channel. The MHTT contains the times when messages have to be transferred by the message transfer process from the *Out* queue into the MBI, in order to be broadcast by the TTP controller. As result of this synchronization, the activation times in the MHTT

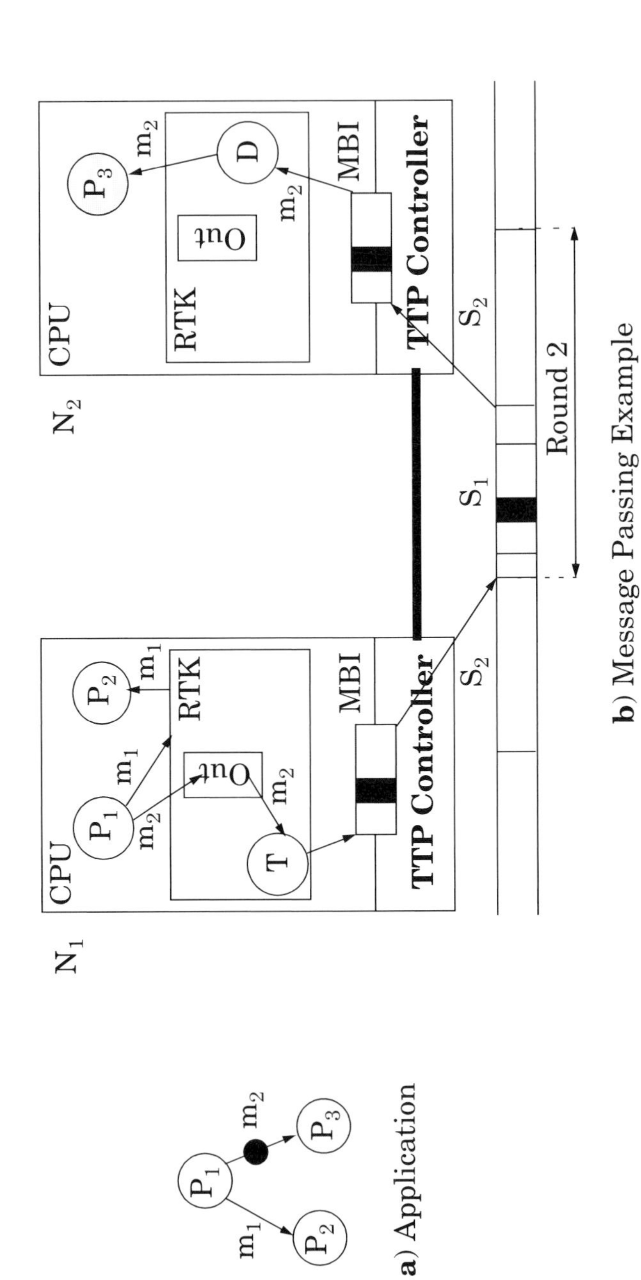

Figure 3.6: A Message Passing Example for Event-Driven Systems

are directly related to those in the MEDL and the first table results directly from the second one.

It is easy to observe that we have the most favorable situation when, at a certain activation, the message transfer process finds in the *Out* queue all the "expected" messages which then can be packed into the next following frame to be sent by the TTP controller. However, application processes are not statically scheduled and availability of messages in the *Out* queue can not be guaranteed at fixed times. Worst case situations have to be considered, as will be shown in Section 6.4.

Let us come back to Figure 3.6. There we assumed a context in which the broadcasting of the frame containing message m_2 is done in the slot S_1 of *Round 2*. The TTP controller of node N_2 knows from its MEDL that it has to read a frame from slot S_1 of *Round 2* and to transfer it into its MBI. In order to synchronize with the TTP controller and to read the frame from the MBI, the scheduler on node N_2 will activate, based on its local MHTT, a so called delivery process, denoted with D in Figure 3.6. The delivery process takes the frame from the MBI and extracts the messages from it. For the case when a message is split into several packets, sent over several TDMA rounds, we consider that a message has arrived at the destination node after all its corresponding packets have arrived. When m_2 has arrived, the delivery process copies it to process P_3 which will be activated. Activation times for the delivery process are fixed in the MHTT just as explained earlier for the message transfer process.

The number of activations of the message transfer and delivery processes depends on the number of frames transferred, and they are taken into account for timing analysis, as well as the delay implied by the propagation on the communication bus.

3.5 Multi-Cluster Systems

A multi-cluster system consists of several clusters, interconnected by gateways (Figure 3.7 depicts a two-cluster example). A *cluster* is composed of nodes which share a broadcast communication channel.

In a *time-triggered cluster* (TTC), processes and messages are scheduled according to a static cyclic policy, with the bus implementing the TTP. On an *event-triggered cluster* (ETC), the processes are scheduled according to a priority based preemptive approach, while messages are transmitted using the priority-based CAN protocol.

The critical element of such an architecture is the gateway, which is a node connected to both types of clusters (hence having two communication controllers, for TTP and CAN), and is responsible for the inter-cluster routing of real-time traffic.

Although in this book we consider only a two cluster system, as the one in Figure 3.7, the approaches presented can be easily

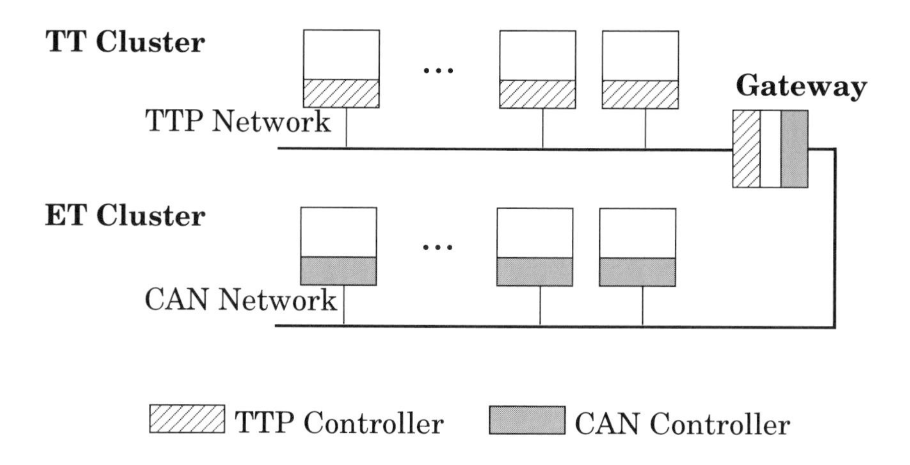

Figure 3.7: A Two-Cluster System Example

extended to cluster configurations where there are several ETCs and TTCs interconnected by gateways.

A real-time kernel is responsible for activation of processes and transmission of messages on each node. On a TTC, the processes are activated based on the local schedule tables, and messages are transmitted according to the MEDL. On an ETC, we have a scheduler that decides on activation of ready processes and transmission of messages, based on their priorities.

In Figure 3.8 we illustrate our message passing mechanism. Here we concentrate on the communication between processes located on different clusters. For message passing details within a TTC the reader is directed to Section 3.3, while the infrastructure needed for communications on an ETC has been detailed in [Tin95].

Let us consider the example in Figure 3.8, where we have an application consisting of four processes, mapped on two clusters. Processes P_1 and P_4 are mapped on node N_1 of the TTC, while P_2 and P_3 are mapped on node N_2 of the ETC. Process P_1 sends messages m_1 and m_2 to processes P_2 and P_3, respectively, while P_2 and P_3 send messages m_3 and m_4 to P_4.

The transmission of messages from the TTC to the ETC takes place in the following way (see Figure 3.8). P_1, which is statically scheduled, is activated according to the schedule table, and when it finishes it calls the send kernel function in order to send m_1 and m_2, indicated in the figure by number (1). Messages m_1 and m_2 have to be sent from node N_1 to node N_2. At a certain time, known from the schedule table, the kernel transfers m_1 and m_2 to the TTP controller by packing them into a frame in the MBI. Later on, the TTP controller knows from its MEDL when it has to take the frame from the MBI, in order to broadcast it on the bus. In our example, the timing information in the schedule table of the kernel and the MEDL is determined in such a way that the broadcasting of the frame is done in the slot S_1 of *Round 2* (2). The TTP controller of the gateway node N_G knows from its MEDL that it has to read a frame from slot S_1 of *Round 2*

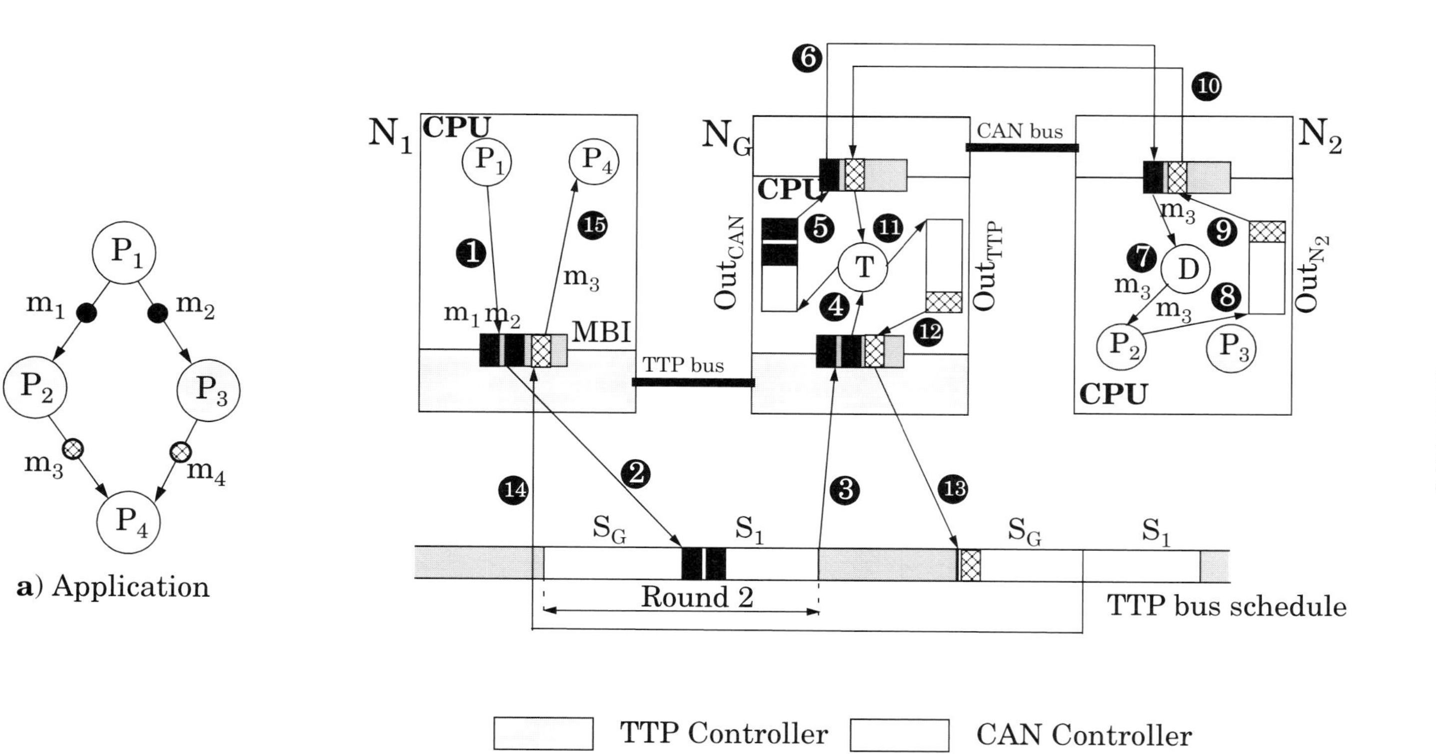

Figure 3.8: A Message Passing Example for Multi-Cluster Systems

and to transfer it into its MBI (3). Invoked periodically, having the highest priority on node N_G, and with a period which guarantees that no messages are lost, the gateway process T copies messages m_1 and m_2 from the MBI to the TTP-to-CAN priority-ordered message queue Out_{CAN} (4). The highest priority message in the queue, in our case m_1, will tentatively be broadcast on the CAN bus (5). Whenever message m_1 will be the highest priority message on the CAN bus, it will successfully be broadcast and will be received by the interested nodes, in our case node N_2 (6). The CAN communication controller of node N_2 receiving m_1 will copy it in the transfer buffer between the controller and the CPU, and raise an interrupt which will activate a delivery process D, responsible to activate the corresponding receiving process, in our case P_2, and hand over message m_1 that finally arrives at the destination (7).

Message m_3 (depicted in Figure 3.8 as a hashed rectangle) sent by process P_2 from the ETC will be transmitted to process P_4 on the TTC. The transmission starts when P_2 calls its send function and enqueues m_3 in the priority-ordered Out_{N_2} queue (8). When m_3 has the highest priority on the bus, it will be removed from the queue (9) and broadcast on the CAN bus (10), arriving at the gateway's CAN controller. The gateway transfer process T is activated, and m_3 is placed in the Out_{TTP} FIFO queue (11). The gateway node N_G is only able to broadcast on the TTC in its corresponding slot S_G of the TDMA rounds circulating on the TTP bus. According to the MEDL of the gateway, a set of messages not exceeding $size_{S_G}$ of the slot S_G will be removed from the front of the Out_{TTP} queue in every round, and packed in the S_G slot (12). Once the frame is broadcast (13) it will arrive at node N_1 (14), where all the messages in the frame will be copied in the input buffers of the destination processes (15). Process P_4 is activated according to the schedule table, which has to be constructed such that it accounts for the worst-case communication delay of messages m_3 and m_4, bounded by the analysis in Section 8.2.1, and, thus, when P_4 starts executing it will find m_3 and m_4 in its input buffer.

This chapter has presented the architectures of the considered distributed systems. In part two of the book, consisting of chapters 4 and 5 we will address time-driven systems, in the third part, chapters 6 and 7, event-driven systems are considered, and in the fourth part, chapters 8, 9 and 10, we discuss issues related to multi-cluster systems.

PART II
Time-Driven Systems

Chapter 4
Scheduling and Bus Access Optimization for Time-Driven Systems

IN THIS AND in the following chapter we consider time-driven distributed real-time systems that use the time-triggered protocol for their communication infrastructure, as described in Section 3.3. In this case, both the activation of processes and the transmission of messages are done based on the progression of time. The applications are modeled as a set of conditional process graphs, as presented in Section 2.3.1.

Our goal in this chapter is to generate a static schedule and to optimize the parameters of the communication protocol, such that the worst-case delay by which the system completes execution is minimized.

The chapter starts by presenting an approach to static scheduling with control and data dependencies for distributed real-time systems [Dob98], [Ele98a], [Ele00]. The approach considers a simplified communication model in which the execution time of the communication processes depends only on the amount of data exchanged by the processes engaged in the communication. The communication processes are treated exactly as ordinary processes during scheduling, and the bus is modeled similar to a programmable processor that can "execute" one communication at a time as soon as the communication becomes "ready".

We propose in this chapter several extensions to this basic approach:

- scheduling of messages using a realistic communication model based on the time-triggered protocol (Section 4.3.1);
- a new priority function for list scheduling that uses knowledge about the bus access scheme in order to improve the schedule quality (Section 4.3.2);
- optimization strategies for the synthesis of parameters of the communication protocol, aimed at improving the schedule quality (Section 4.4).

4.1 Background

Static cyclic scheduling of a set of data dependent software processes on a multiprocessor architecture has been intensively researched [Kop97a], [Xu00].

Several approaches are based on list scheduling heuristics using different priority criteria [Cof72], [Deo98], [Jor97], [Kwo96], [Wu90] or on branch-and-bound algorithms [Kas84]. These approaches are based on the assumption that a number of identical processors are available to which processes are progressively assigned as the static schedule is elaborated. Such an assumption is obviously not acceptable for distributed embedded systems which are heterogeneous by nature. In [Jor97] a list scheduling based approach is extended to handle heterogeneous architectures. Scheduling is performed by progressively assigning tasks to the allocated processors with the goal to minimize the length of the schedule. The proposed algorithm handles only processors which execute one single process at a time (not typical for hardware) and the resulting partitioning does not take into consideration any design constraints.

In [Ben96], [Pra92] static scheduling and partitioning of processes, and allocation of system components, are formulated as a mixed integer linear programming (MILP) problem. A disadvantage of this approach is the complexity of solving the MILP model. The size of such a model grows quickly with the number of processes and allocated resources. In [Kuc97] a formulation using constraint logic programming has been proposed for similar problems.

In all the previous approaches process interaction is only in terms of dataflow. However, when including control dependencies significant improvements in the quality of the resulting schedules can be obtained [Ele98a], [Ele00], [Kuc01]. Section 4.2 presents in more detail related research on the static scheduling for systems with control and data dependencies that is used as a starting point for our work.

It has been claimed [Xu93] that static cyclic scheduling is the only approach that can provide solutions to applications that exhibit data dependencies. However, advances in the area of fixed priority preemptive scheduling show that such applications can also be handled with other scheduling strategies [Aud93], [Tin94b], [Pal98], [Pal99].

Currently, more and more real-time systems are used in physically distributed environments and have to be implemented on distributed architectures in order to meet reliability, functional, and performance constraints. However, researchers have often ignored or very much simplified aspects concerning the communication infrastructure.

One typical approach is to consider communication processes as processes with a given execution time (depending on the amount of information exchanged) and to schedule them as any other process, without considering issues like communication protocol, bus arbitration, packing of messages, clock synchronization, etc. These aspects are, however, essential in the context of safety-critical distributed real-time applications and one of our objectives is to develop a strategy which takes them into consideration for process scheduling.

Many efforts dedicated to communication synthesis have concentrated on the synthesis support for the communication infrastructure but without considering hard real-time constraints and system level scheduling aspects [Cho95b], [Dav95], [Knu99], [Nar94]. Lower level communication synthesis aspects under timing constraints have been addressed in [Ort98], [Knu99].

4.2 Scheduling with Control and Data Dependencies

The problem which is discussed in this section can be formulated as follows: Given an application distributed on a time-driven system (Section 3.3), modeled as a set of mapped conditional process graphs (Section 2.3.1), we are interested to generated a static schedule such that the worst-case delay by which the system completes execution is minimized.

According to our application model, some processes can only be activated if certain conditions, computed by previously executed processes, are fulfilled. Hence, process scheduling is complicated since at a given activation of the system, only a certain subset of the total amount of processes is executed and this subset differs from one activation to the other.

As the values of the conditions are unpredictable, the decision on which process to activate and at which time has to be taken without knowing which values the conditions will later get. On the other side, at a certain moment during execution, when the values of some conditions are already known, they have to be used in order to take the best possible decisions on when and which process to activate. Heuristic algorithms have to be developed to produce a schedule of the processes such that the worst case delay is as small as possible. One such algorithm will be presented in Section 4.2.1.

The output produced by their scheduling algorithm is a schedule table that contains all the information needed by a distributed run time scheduler to take decisions on activation of processes. It is considered that, during execution, a very simple non-preemptive scheduler located in each processing element decides on process and communication activation depending on the actual values of conditions. Only one part of the table has to be stored in each processor, namely, the part concerning decisions which are taken by the corresponding scheduler.

Example 4.1: Under these assumptions, Table 4.1 presents a possible schedule (produced by the algorithm in Figure 4.1) for the conditional process graph in Figure 2.5 on page 29. In Table 4.1 there is one row for each "ordinary" or communication process, which contains activation times corresponding to different values of conditions. Each column in the table is headed by a logical expression constructed as a conjunction of condition values. Activation times in a given column represent starting times of the processes when the respective expression is true.

According to the schedule in Table 4.1 process P_1 is activated unconditionally at the time 0, given in the first column of the table. Activation of the rest of the processes, in a certain execution cycle, depends on the values of the conditions, which are unpredictable. For example, process P_{11} has to be activated at $t = 44$ if $C \wedge D$ is true and at $t = 52$ if $\overline{C} \wedge D$ holds. ■

At a certain moment during the execution, when the values of some conditions are already known, they have to be used in order to take the best possible decisions on when and which process to activate. Therefore, after the termination of a process that produces a condition (disjunction process), the value of the condition is broadcast from the corresponding processor to all

Table 4.1: Schedule Table for the Process Graph in Figure 2.5

Process	True	C	$C \wedge D$	$C \wedge \overline{D}$	$\overline{C}$	$\overline{C} \wedge D$	$\overline{C} \wedge \overline{D}$
P_1	0						
P_2		5					
P_3			14	14			
P_4			45	45			
P_5			51	50		55	47
P_6		3			3		
P_7		7			7		
P_8			9			9	
P_9			11			11	
P_{10}			13			13	
P_{11}			44			52	
P_{12}			47	9		55	9
P_{13}			48	13		56	11
P_{14}						14	9
$P_{1,2}$		4					
$P_{4,5}$			48	47			
$P_{2,3}$			13	13			
$P_{3,4}$			44	44			
$P_{12,13}$			47	10		55	
$P_{8,10}$			12			12	
$P_{10,11}$			43			43	
C		3				11	9
D			11	9		11	9

other processors. This broadcast is scheduled as soon as possible on the communication channel, and is considered together with the scheduling of the messages.

To produce a deterministic behavior, which is correct for any combination of conditions, the table has to fulfill several requirements:

1. No process will be activated if, for a given execution, the conditions required for its activation are not fulfilled.
2. Activation times have to be uniquely determined by the conditions.
3. Activation of a process P_i at a certain time t has to depend only on condition values which are determined at the respective moment t and are known to the processing element which executes P_i.

4.2.1 LIST SCHEDULING BASED ALGORITHM

Optimal scheduling has been proven to be an NP-complete problem [Ull75] in even simpler contexts than those characteristic to distributed systems represented as CPGs. Hence, it is essential to develop heuristics which produce good quality results in a reasonable time.

In [Dob98], [Ele98a], [Ele00] the authors concentrate on developing a scheduling algorithm for systems with both control and data dependencies, modeled using the conditional process graph. As the starting point for our improved scheduling technique that is tailored for time-triggered embedded systems we consider the list scheduling based algorithm in [Dob98], [Ele00] presented, in a simplified form, in Figure 4.1.

List scheduling heuristics [Ele98b], [Ele00] are based on priority lists from which processes are extracted in order to be scheduled at certain moments. In the algorithm presented in Figure 4.1, there is such a list, ReadyList, which contains the processes eligible to be activated on the corresponding processor at time CurrentTime. These are processes which have not yet been scheduled but have all predecessors already scheduled and terminated.

The ListScheduling function is recursive and calls itself for each disjunction node in order to separately schedule the nodes in the true branch, and those in the false branch, respectively (lines 10 and 13 in Figure 4.1). Thus, the alternative paths are not activated simultaneously and resource sharing is correctly achieved (for details on how the algorithm fulfills the three requirements on the schedule table, identified earlier, we refer to [Ele00]).

An essential component of a list scheduling heuristic is the priority function used to solve conflicts between ready processes. As can be observed in Figure 4.1, the highest priority process

```
ListScheduling(CurrentTime, ReadyList, KnownConditions)
1  repeat
2      Update(ReadyList)
3      for each processing element PE do
4          if PE is free at CurrentTime then
5              P_i = GetReadyProcess(ReadyList)
6              if there exists a P_i then
7                  Insert(P_i, ScheduleTable, CurrentTime, KnownConds)
8                  if P_i is a disjunction process then
9                      C_i = condition calculated by P_i
10                     ListScheduling(CurrentTime,
11                         ReadyList ∪ ready nodes from the true branch,
12                         KnownConditions ∪ true C_i)
13                     ListScheduling(CurrentTime,
14                         ReadyList ∪ ready nodes from the false branch,
15                         KnownConditions ∪ false C_i)
16                 end if
17             end if
18         end if
19     end for
20     CurrentTime = earliest time when a scheduled process terminates
21 until all processes of this alternative path are scheduled
end ListScheduling
```

Figure 4.1: List Scheduling Based Algorithm for CPGs

will be extracted by function GetReadyProcess from the ReadyList in order to be scheduled (line 5).

4.2.2 PCP PRIORITY FUNCTION

Priorities for list scheduling very often are based on the critical path (CP) from the respective process to the sink node. Thus, for CP scheduling, the priority assigned to a process P_i will be the maximal execution time from the current node to the sink:

$$l_{P_i} = \max_k \sum_{P_j \in \pi_{ik}} C_{Pj}, \tag{4.1}$$

where π_{ik} is the kth path from node P_i to the sink node.

Considering the particularities of our problem, significant improvements of the resulting schedule can be obtained, without any penalty in scheduling time, by making use of the available information on process allocation [Ele98b].

Let us consider the graph in Figure 4.2 and suppose that the list scheduling algorithm has to decide between scheduling process P_A or P_B which are both ready to be scheduled on the same programmable processor or bus pe_i. In Figure 4.2 we depicted only the critical path from P_A and P_B to the sink node. Let us consider that P_X is the last successor of P_A on the critical path such that all processes from P_A to P_X are assigned to the same processing element pe_i. The same holds for P_Y relative to P_B. Times t_A and t_B are the total execution time of the chain of processes from P_A to P_X and from P_B to P_Y, respectively, following the critical paths. Times λ_A and λ_B are the total execution times of the processes on the rest of the two critical paths. Thus, considering Equation 4.1 we have the following critical paths for P_A and P_B, respectively:

$$l_{P_A} = t_A + \lambda_A, \; l_{P_B} = t_B + \lambda_B.$$

However, the algorithm proposed in [Ele98b] does not use the length of these critical paths as a priority. The policy in [Ele98b]

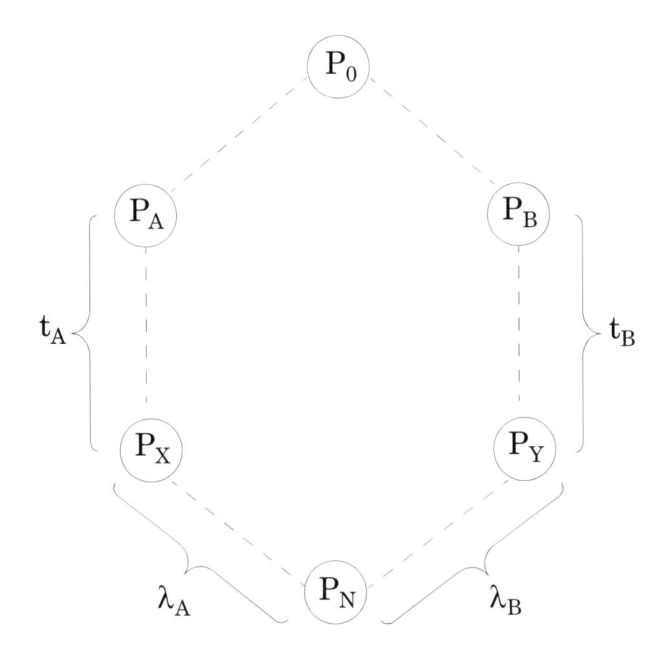

Figure 4.2: Delay Estimation for PCP Scheduling

is based on the estimation of a lower bound L on the total delay, taking into consideration that the two chains of processes $P_A - P_X$ and $P_B - P_Y$ are executed on the same processor. L_{P_A} and L_{P_B} are the lower bounds on the delay if P_A and P_B, respectively, are scheduled first:

$$L_{P_A} = max(T\_current + t_A + \lambda_A, T\_current + t_A + t_B + \lambda_B),$$

$$L_{P_B} = max(T\_current + t_B + \lambda_B, T\_current + t_B + t_A + \lambda_A).$$

The alternative that offers the perspective of the shorter delay $L = \min(L_{P_A}, L_{P_B})$ is selected. It can be observed that if $\lambda_A > \lambda_B$ then $L_{P_A} < L_{P_B}$, which means that we have to schedule P_A first so that $L = L_{PA}$; similarly if $\lambda_B > \lambda_A$ then $L_{PB} < L_{P_A}$, and we have to schedule P_B first in order to get $L = L_{P_B}$.

4.3 Scheduling for Time-Driven Systems

In the previous sections we were interested to derive a static schedule table such that the worst-case delay of an application, modeled as conditional process graphs, is minimized. In this section, we propose several extensions to the scheduling algorithm briefly described in Section 4.2. The extensions consider a realistic communication and execution infrastructure, and include aspects of the communication protocol in the optimization process.

As an input to our problem we consider a safety-critical application modeled as a set of conditional process graphs, see Section 2.3.1. The architecture of the system is given as described in Section 3.3. Each process of the application is mapped on a processor. The worst-case execution time for each process is known, as well as the length S_{m_i} of each message.

We are interested to derive the worst case delay on the system execution time, so that this delay is as small as possible, and to synthesize the local schedule tables for each node, as well as the MEDL for the TTP controllers, which guarantee this delay.

Considering the concrete definition of our problem, which takes into account the details of the communication protocol, the communication time is no longer dependent only on the length of the message, as assumed in the previous section. Hence, if the message is sent between two processes mapped onto different nodes, the message has to be scheduled according to the TTP protocol. Several messages can be packaged together in the data field of a frame. The number of messages that can be packed depends on the slot length corresponding to the node. The effective time spent by a message m_i *on the bus* is $C_{m_i} = b_{S_i} / s$, where S_{S_i} is the length of the slot S_i and s is the transmission speed of the channel. Therefore, the communication time C_{m_i} does not depend on the bit length S_{m_i} of the message m_i, but on the slot length corresponding to the node sending m_i.

Example 4.2: The important impact of the communication parameters on the performance of the application is illustrated in Figure 4.3 by means of a simple example. In Figure 4.3d we have a process graph consisting of four processes P_1 to P_4 and four messages m_1 to m_4. The architecture consists of two nodes interconnected by a TTP channel. The first node N_1 transmits on the slot S_1 of the TDMA round and the second node N_2 transmits on the slot S_2. Processes P_1 and P_4 are mapped on node N_1, while processes P_2 and P_3 are mapped on node N_2.

With the TDMA configuration in Figure 4.3a, where the slot S_2 is scheduled first and slot S_1 is second, we have a resulting schedule length of 24 ms. However, if we swap the two slots inside the TDMA round without changing their lengths, we can improve the schedule by 2 ms, as seen on Figure 4.3b.

Furthermore, if we have the TDMA configuration in Figure 4.3c where slot S_1 is first, slot S_2 is second and we increase the slot lengths so that the slots can accommodate both of the messages generated on the same node, we obtain a schedule length of 20 ms which is optimal.

However, increasing the length of slots does not necessarily improve a schedule, as it delays the communication of messages generated by other nodes.

■

In the next two sections our goal is to synthesize the local schedule table of each node and the MEDL of the TTP controller for a given order of slots in the TDMA round and given slot lengths. The ordering of slots and the optimization of slot lengths will be discussed in Section 4.4.

4.3.1 Scheduling of Messages with the TTP

Given a certain bus access scheme, which means a given ordering of the slots in the TDMA round and fixed slot lengths, a CPG has to be scheduled with the goal to minimize the worst case

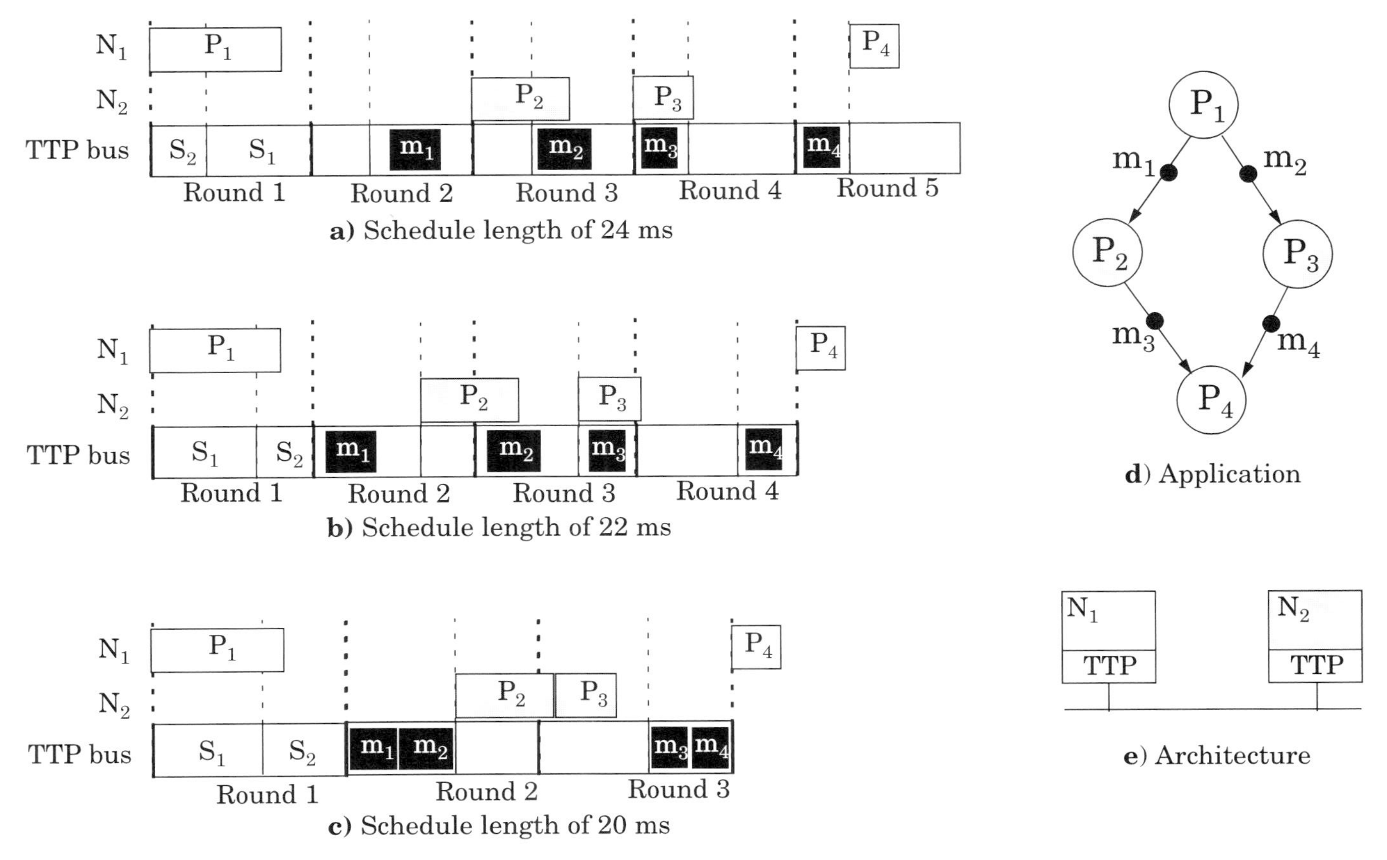

Figure 4.3: Static Cyclic Scheduling Examples with the TTP

execution delay. This can be performed using the algorithm ListScheduling (Figure 4.1) presented in Section 4.2.1. Two aspects have to be discussed here: the planning of messages in predetermined slots and the impact of this communication strategy on the priority assignment.

The function ScheduleMessage in Figure 4.4 is called in order to plan the communication of a message m, with length S_m, generated on $Node_m$ and which is ready to be transmitted at TimeReady. The ScheduleMessage function is called immediately following line five in Figure 4.1, considering the processing element PE as the bus, P_i as the message m (produced with a corresponding GetReadyMessage), and with TimeReady = CurrentTime.

ScheduleMessage returns the earliest round and the corresponding slot (the slot corresponding to $Node_m$) which can host the message. In Figure 4.4 RoundLength is the length of a TDMA round expressed in time units (in Figure 4.5, for example, RoundLength = 18 ms). The first round after TimeReady is the ini-

```
ScheduleMessage (TimeReady, S_m, Node_m)
1   -- the slot in which the message has to be sent
2   Slot=the slot assigned to Node_m
3   -- the first round which could be a candidate
4   Round = ⌈TimeReady / RoundLength⌉
5   -- is the right slot in this round already gone?
6   if TimeReady – Round * RoundLength > start_Slot then
7       -- if yes, take the next round
8       Round = Round + 1
9   end if
10  -- is enough space left in the slot for the message?
11  while S_m > S_Slot – S_occupied do
12      -- if not, take the next round
13      Round = Round + 1
14  end while
15  -- return the right round and slot
16  return (Round, Slot)
end ScheduleMessage
```

Figure 4.4: The ScheduleMessage Function

tial candidate to be considered (line 4). For this round, however, it can be too late to catch the corresponding slot, in which case the next round is selected, lines 5–9. When a candidate round is selected we have to check, in line 11, that there is enough space left in the slot for our message ($S_{occupied}$ represents the total number of bits occupied by messages already scheduled in the respective slot of that round). If no space is left, the communication has to be delayed for another round (line 13).

With this message scheduling scheme, the algorithm in Figure 4.1 will generate correct schedules for a TTP based architecture, with guaranteed worst-case execution delays. However, the quality of the schedules can be much improved by adapting the priority assignment scheme so that particularities of the communication protocol are taken into consideration.

4.3.2 IMPROVED PRIORITY FUNCTION

For the scheduling algorithm outlined previously we initially used the Partial Critical Path (PCP) priority function presented in Section 4.2.2. As discussed before, PCP uses as a priority criterion the length of that part of the critical path corresponding to a process P_i which starts with the first successor of P_i that is assigned to a processor different from the processor running P_i. The PCP priority function is statically computed once at the beginning of the scheduling procedure.

However, considering the concrete definition of our problem, significant improvements of the resulting schedule can be obtained by including knowledge of the bus access scheme into the priority function. This new priority function will be used by the GetReadyProcess (Figure 4.1) in order to decide which process to select from the list of ready process.

Example 4.3: Let us consider the graph in Figure 4.5c, and suppose that the list scheduling algorithm has to decide whether to schedule process P_1 or P_2 which are both ready to be scheduled on the same programmable processor. The

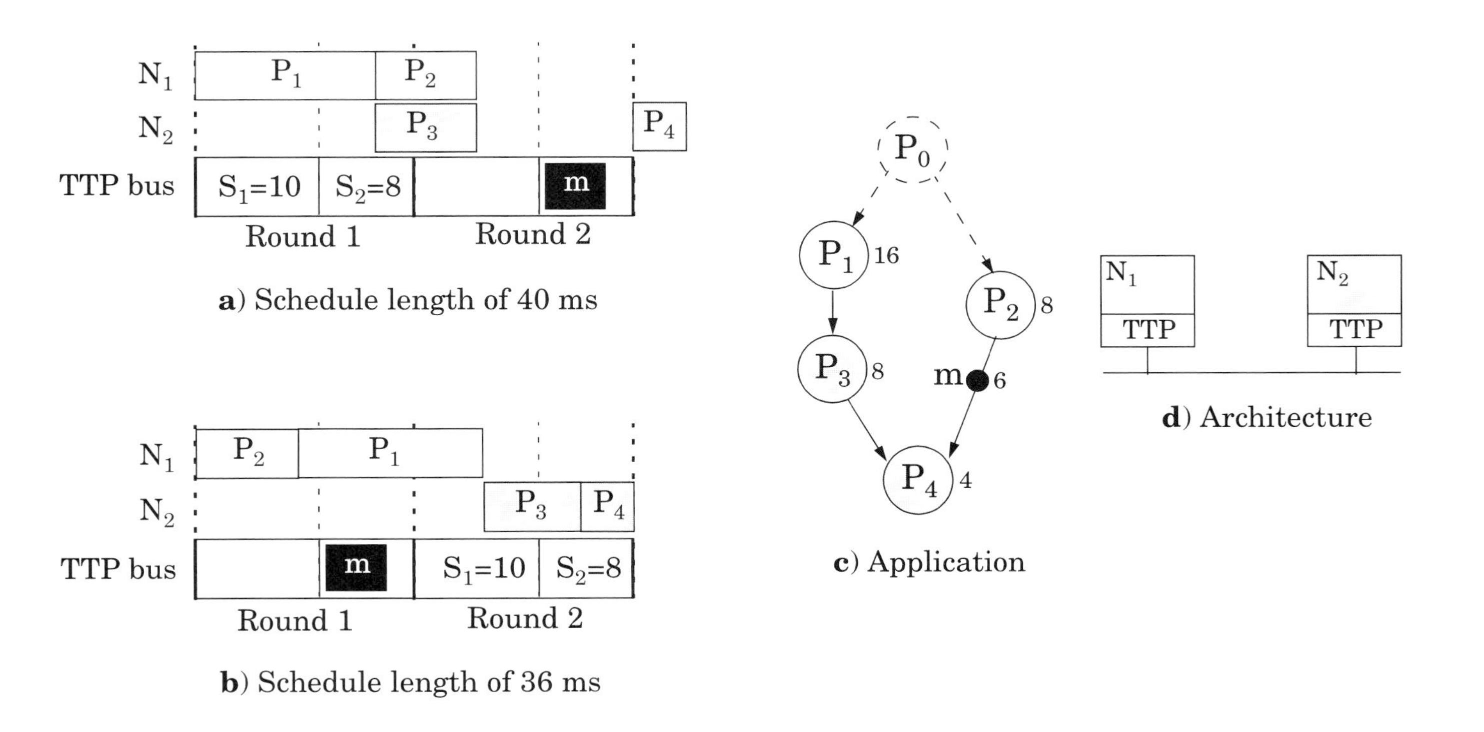

a) Schedule length of 40 ms

b) Schedule length of 36 ms

c) Application

d) Architecture

Figure 4.5: Priority Function Example

worst-case execution time of the processes is depicted on the right side of the respective node and is expressed in ms. The architecture consists of two nodes interconnected by a TTP channel. Processes P_1 and P_2 are mapped on node N_1, while processes P_3 and P_4 are mapped on node N_2. Node N_1 transmits in slot S_1 of the TDMA round and N_2 transmits in slot S_2. Slot S_1 has a length of 10 ms while slot S_2 has a length of 8 ms. For simplicity we suppose that there is no message transferred between P_1 and P_3. The PCP (see Section 4.1.2) assigns a higher priority to P_1 because it has a partial critical path of 12, starting from P_3, longer than the partial critical path of P_2 which is 10 and starts from m. This results in a schedule length of 40 ms as depicted in Figure 4.5a. On the other hand, if we schedule P_2 first, the resulting schedule, depicted in Figure 4.5b, is of only 36 ms.

This apparent anomaly is due to the fact that the way we have computed PCP priorities, considering message communication as a simple activity of delay 6ms, is not realistic in the context of a TDMA protocol. Let us consider the particular TDMA configuration in Figure 4.4 and suppose that the scheduler has to decide at $t = 0$, which one of the processes P_1 or P_2 to schedule. If P_2 is scheduled, the message is ready to be transmitted at $t' = 8$. Based on a computation similar to that used in Figure 4.5, it follows that message m will be placed in round $\lceil 8/18 \rceil$[1] = 1, and it arrives in time to get slot S_1 of that round ($\text{TimeReady} = 8 < \text{start}_{S_1} = 10$). Thus, m arrives at $t_{arr} = 18$, which means a delay relative to $t' = 8$ (when the message was ready) of $\delta = 10$. This is the delay that should be considered for computing the partial critical path of P_2, which now results in $\delta + t_{P_4} = 14$ (longer than the one corresponding to P_1).

■

1. The operator $\lceil x \rceil$ is the ceiling operator, which returns the smallest integer greater than or equal to x.

The obvious conclusion is that priority estimation has to be based on message planning with the TDMA scheme. Such an estimation, however, cannot be performed statically, before scheduling. If we take the same example in Figure 4.5, but consider that the priority based decision is taken by the scheduler at $t = 5$, m will be ready at $t' = 13$. This is too late for m to get into slot S_1 of *Round 1*. The message arrives with *Round 2* at $t_{arr} = 36$. This leads to a delay due to the message passing of $\delta = 36 - 13 = 23$, different from the one computed above.

We introduce, therefore, a new priority function, the Modified PCP (MPCP), which is computed during scheduling, whenever several processes are in competition to be scheduled on the same resource. Similar to PCP, the priority metric is the length of that portion of the critical path corresponding to a process P_i which starts with the first successor of P_i that is assigned to a processor different from $M(P_i)$. The critical path estimation starts with time t at which the processes in competition are ready to be scheduled on the available resource. During the partial traversal of the graph the delay introduced by a certain node P_j is estimated as follows:

$$\delta_{Pj} = \begin{cases} t_{Pj}, \text{ if } P_j \text{ is not a message passing} \\ t_{arr} - t', \text{ if } P_j \text{ is a message passing} \end{cases}$$

The term t' is the time when the node generating the message terminates (and the message is ready); t_{arr} is the time when the slot to which the message is supposed to be assigned has arrived. The slot is determined like in Figure 4.4, but without taking into consideration space limitations in slots.

Thus, the priority function MPCP has to be dynamically determined during the scheduling algorithm for each ready process, every time the GetReadyProcess function is activated in order to select a process from the ReadyList. The computation of λ, used in MPCP similarly to the PCP case (see Section 4.2.2), is performed inside the GetReadyProcess function and involves a partial traversal of the graph, as presented in Figure 4.6.

As the experimental results (Section 4.5) show, using MPCP instead of PCP for the TTP based architecture results in an important improvement of the quality of generated schedules, only with a slight increase in scheduling time.

4.4 Bus Access Optimization

In the previous sections we have shown how the algorithm ListScheduling can produce an efficient schedule for a CPG, given a certain TDMA bus access scheme. However, as shown in

```
Lambda(lambda, CurrentProcess)
1  if CurrentProcess is a message then
2      slot = slot of node sending CurrentProcess
3      round = lambda / RoundLength
4      if lambda – RoundLength * round > start of slot in round then
5          round = next round
6      end if
7      while not message fits in the slot of round do
8          round = next round
9      end while
10     lambda = round * RoundLength + start of slot in round + length of slot
11 else
12     lambda = lambda + WCET of CurrentProcess
13 end if
14 if lambda > MaxLambda then
15     MaxLambda = lambda
16 end if
17 for each successor of CurrentProcess do
18     Lambda(lambda, successor)
19 end for
20 return MaxLambda
end Lambda
```

Figure 4.6: The Lambda Function

Figure 4.3 on page 77, both the ordering of slots and the slot lengths strongly influence the worst-case execution delay of the system.

In this section, we first present a heuristic which, based on a greedy approach, determines an ordering of slots and their lengths so that the worst-case delay corresponding to a certain CPG is as small as possible. Then, we present an algorithm based on a simulated annealing strategy, which finds that bus configuration which leads to the near-optimal delay for a CPG.

4.4.1 GREEDY APPROACHES

Figure 4.7 presents a greedy heuristic that starts with determining an initial solution, the so called "straightforward" one, which assigns in order nodes to the slots ($Node_{S_i} = N_i$) and fixes the slot length $length_{S_i}$ to the minimal allowed value, which is equal to the length of the largest message generated by a process assigned to $Node_{S_i}$ (lines 1–5).

The next step of the algorithm starts with the first slot and tries to find the node which, when transmitting in this slot, will minimize the worst case delay of the system, as produced by ListScheduling. Simultaneously with searching for the right node to be assigned to the slot, the algorithm looks for the optimal slot length (lines 12–18). Once a node was selected for the first slot and a slot length fixed (line 23), the algorithm continues with the next slots, trying to assign nodes (and to fix slot lengths) from those nodes which have not yet been assigned.

When calculating the length of a certain slot, a first alternative could be to try all the slot lengths S_S allowed by the protocol. Such an approach starts with the minimum slot length determined by the largest message to be sent from the candidate node, and it continues incrementing with the smallest data unit (e.g., 2 bits) up to the largest slot length determined by the maximum allowed data field in a TTP frame (e.g., 32 bits, depending on the controller implementation). We call this alternative OptimizeAccess1.

A second alternative, OptimizeAccess2, is based on a feedback from the scheduling algorithm which recommends slot sizes to be tried out. Before starting the actual optimization process for the bus access scheme, a scheduling of the straightforward solution (determined in lines 1–5) is performed which generates the recommended slot lengths. These lengths are produced by the ScheduleMessage function (Figure 4.4), whenever a new round has to be selected because of lack of space in the current slot. In such a case the slot length which would be needed in order to accommodate the new message is added to the list of recom-

```
OptimizeAccess
1  -- creates the initial, straightforward solution
2  for i = 1 to NrSlot do
3      Node_S = N_i
4      length_S = MinLength_Si
5  end for
6  -- over all slots
7  for i = 1 to NrSlot do
8      -- over all slots which have not yet been allocated
9      -- a node and slot length
10     for j = i to NrSlot do
11         swap values (Node_Si, length_Si) with (Node_Sj, length_Sj)
12         -- initially, length_Si has the minimal allowed value
13         for all slot lengths S_S, larger than length_Si do
14             length_S = S_S
15             ListScheduling( ... )
16             remember BestSolution = (Node_Si, length_Si),
17                 with the smallest δ_max produced by ListScheduling
18         end for
19         swap back values (Node_Si, length_Si) with (Node_Sj, length_Sj)
20             to the state before entering the for cycle
21     end for
22     -- slot S_i gets a node allocated and a length fixed
23     Bind (Node_Si, length_Si) = BestSolution
24 end for
end OptimizeAccess
```

Figure 4.7: Optimization of the Bus Access Scheme

mended lengths for the respective slot. With this alternative, the optimization algorithm in Figure 4.7 only selects among the recommended lengths when searching for the right dimension of a certain slot (line 13).

4.4.2 SIMULATED ANNEALING

The second algorithm we have developed is based on a simulated annealing (SA) strategy, described in detail in Appendix A.

The greedy strategy constructs the solution by progressively selecting the best candidate in terms of the schedule length produced by the function ListScheduling. Unlike the greedy strategy, SA will try to escape from a local optimum by randomly choosing a neighboring solution, see Figure A.1 on page 286 in Appendix A.

The neighbors of the current solution are obtained by a permutation of the slots in the TDMA round and/or by increasing/decreasing the slot lengths. We generate the new solution by either randomly swapping two slots (with a probability 0.3) and/or by increasing/decreasing with the smallest data unit the length of a randomly selected slot (with a probability 0.7). These probabilities have been determined experimentally.

For graphs with 160 and less processes we were able to run an exhaustive search that found the optimal solutions. For the rest of the graph dimensions, we performed very long and expensive runs with the SA algorithm, and the best solution ever produced has been considered as the optimum for the further experiments. Based on further experiments we have determined the parameters of the SA algorithm so that the optimization time is reduced as much as possible but the optimal result is still produced (see Appendix A for the details on these parameters). For example, for the graphs with 320 nodes, the initial temperature *TI* is 500, the temperature length parameter *TL* is 400 and the cooling ratio ε is 0.97. The algorithm stops if for three consecutive temperatures no new solution has been accepted.

4.5 Experimental Evaluation

For the evaluation of our scheduling algorithms we first used conditional process graphs generated for experimental purpose. We considered architectures consisting of 2, 4, 6, 8 or 10 nodes. Forty processes were assigned to each node, resulting in applications of 80, 160, 240, 320 or 400 processes. Thirty applications were generated for each dimension, thus a total of 150 applications were used for the experimental evaluation. Execution times and message lengths were assigned randomly using both uniform and exponential distribution. For the communication channel we considered a transmission speed of 256 Kbps and a length below 20 meters. The maximum length of the data field was 8 bytes, and the frequency of the TTP controller was chosen to be 20 MHz. All experiments were run on a SPARCstation 20.

4.5.1 Priorities for the TTP Scheduling

The first result concerns the quality of the schedules produced by the list scheduling based algorithm using the PCP and the MPCP priority functions. In order to compare the two priority functions, we have calculated the average percentage deviations of the schedule length produced with PCP and MPCP from the length of the best schedule between the two. The results are depicted in Figure 4.8a. In average the deviation with MPCP is 11.34 times smaller than with PCP. However, due to its dynamic nature, MPCP has in average a bigger execution time than PCP. The average execution times for the ListScheduling function using PCP and MPCP are depicted in Figure 4.8b and are under half a second for graphs with 400 processes.

4.5.2 Bus Access Optimization Heuristics

In the next experiments we were interested to check the potential of the algorithms presented in Section 4.4 to improve the generated schedules by optimizing the bus access scheme. We

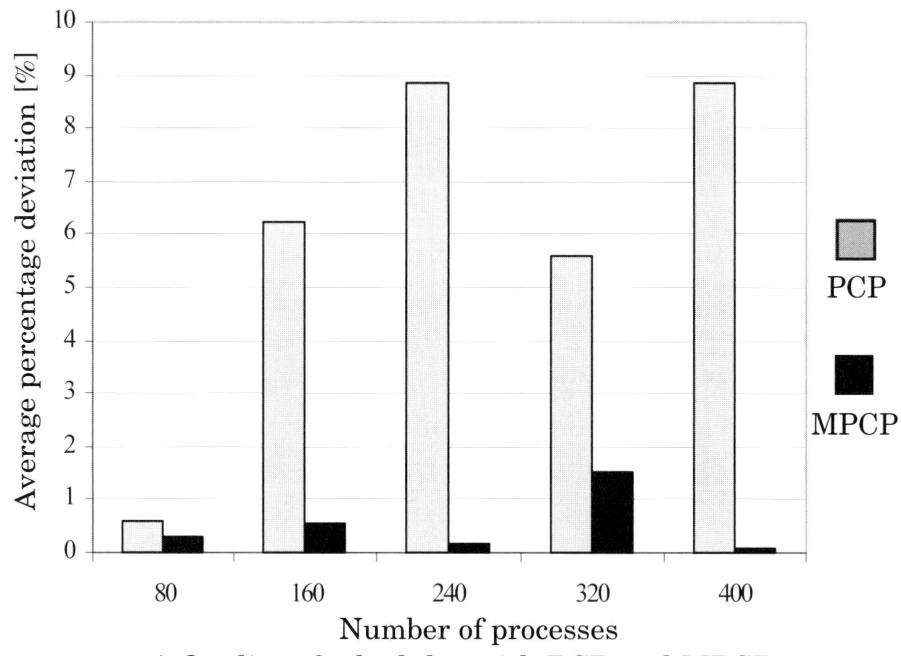

a) Quality of schedules with PCP and MPCP

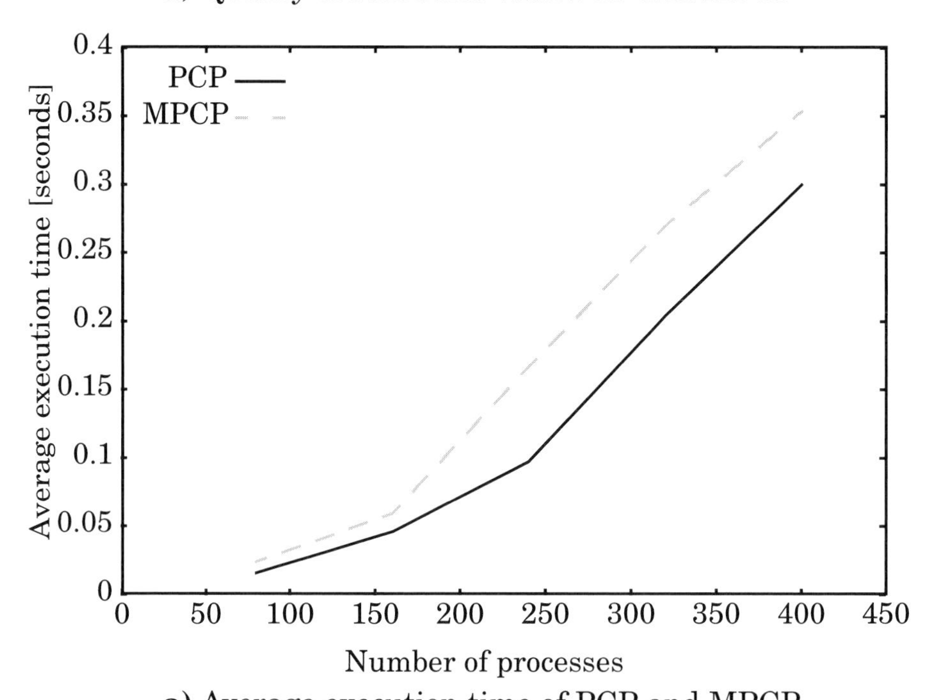

a) Average execution time of PCP and MPCP

Figure 4.8: Comparison of the Two Priority Functions

compared schedule lengths, obtained for the 150 applications in the previous section, considering four different bus access schemes: the straightforward solution, the optimized schemes generated with the two alternatives of our greedy algorithm (OptimizeAccess1 and OptimizeAccess2) and the near-optimal scheme produced using the simulated annealing (SA) based algorithm. Very long and extensive runs have been performed with the SA algorithm for each application and the best ever solution produced has been considered as the near-optimum for that case.

Table 4.2 presents the average and maximum percentage deviation of the schedule lengths obtained with the straightforward solution and with the two optimized schemes from the length obtained with the near-optimal scheme. For each of the application dimensions, the average optimization time, expressed in seconds, is also given.

The first conclusion is that by considering the optimization of the bus access scheme, the results improve significantly compared to the straightforward solution. The greedy heuristic performs well for all the graph dimensions. As expected, the alternative OptimizeAccess1 (which considers all allowed slot lengths) produces slightly better results, on average, than

Table 4.2: Evaluation of the Bus Access Optimization Algorithms

No. of proc.	Straightforward solution		OptimizeAccess1			OptimizeAccess2		
	avg. dev.	max. dev.	avg. dev.	max. dev.	exec. time	avg. dev.	max. dev.	exec. time
80	3.16%	21%	0.02%	0.5%	0.25s	1.8%	19.7%	0.04s
160	14.4%	53.4%	2.5%	9.5%	2.07s	4.9%	26.3%	0.28s
240	37.6%	110%	7.4%	24.8%	10.46s	9.3%	31.4%	1.34s
320	51.5%	135%	8.5%	31.9%	34.69s	12.1%	37.1%	4.8s
400	48%	135%	10.5%	32.9%	56.04s	11.8%	31.6%	8.2s

OptimizeAccess2. However, the execution times are much smaller for OptimizeAccess2. It is interesting to mention that the average execution times for the SA algorithm, needed to find the near-optimal solutions, are between 5 minutes for the applications with 80 processes and 275 minutes for 400 processes.

4.5.3 The Vehicle Cruise Controller

Finally, we have evaluated our approaches using the cruise controller case study presented in Section 2.3.3. For the implementation of the cruise controller as a time-driven system we have considered:

- the hardware architecture from Figure 2.7a on page 37, consisting of five nodes interconnected by a TTP bus,
- the software architecture for time-triggered systems, outlined in Section 3.3,
- the mapped model presented in Figure 2.9 on page 40, having 32 processes and two conditions,
- and a deadline of 400 ms.

Thus, for the cruise controller example, the straightforward solution for bus access resulted in a schedule corresponding to a maximal delay of 429 ms (which does not meet the deadline) when PCP was used as a priority function, while using MPCP we obtained a schedule length of 398 ms. The first and second greedy heuristics for bus access optimization produced solutions that reduced the worst-case delay to 314 and 323 ms, respectively. The near-optimal solution (produced with the SA based approach) results in a delay of 302 ms. The greedy heuristics and the SA have used MPCP as the priority function for list scheduling.

This shows that the quality of generated schedules can be improved by considering the exact details of the communication protocol, and by optimizing the bus access scheme.

Using as a basis the timing analysis and communication synthesis developed in this chapter, in the next chapter we will address the mapping design task within an incremental design environment.

Chapter 5
Incremental Mapping for Time-Driven Systems

IN THIS CHAPTER we present an approach to mapping and scheduling for time-driven systems where processes are scheduled according to a non-preemptive static cyclic scheduling scheme, and communication uses a time division multiple access (TDMA) protocol. We accurately take into consideration the communication costs and consider, during the mapping and scheduling process, the particular requirements of the communication protocol.

The mapping and scheduling tasks are considered in the context of an incremental design process as outlined in Section 2.2. This implies that we perform mapping and scheduling of new functionality on a given distributed embedded system, so that certain design constraints are satisfied and, in addition:

1. The already running applications are disturbed as little as possible.
2. There is a good chance that, later, new functionality can easily be mapped on the resulted system.

We propose a new heuristic, together with the corresponding design criteria, which finds the set of old applications that have to be re-mapped and rescheduled at the same time with mapping and scheduling the new application, such that the disturbance on the running system (expressed as the total cost implied by the modifications) is minimized. Once this set of applications has been determined, mapping and scheduling are performed according to the requirements stated above.

Supporting such a design process is of critical importance for current and future industrial practice, as the time interval between successive generations of a product is continuously decreasing, while the complexity due to increased sophistication of new functionality is growing rapidly. The goal of reducing the overall cost of successive product generations has been one of the main motors behind the, currently very popular, concept of platform-based design (see Section 2.1.4).

Addressing mapping and scheduling inside an incremental design process is not limited to time-driven systems. In Chapter 7 we investigate the issues arising from considering incremental mapping and scheduling in the context of event-driven systems, where processes are scheduled according to a fixed-priority preemptive scheme, while messages are sent using the TTP.

For the sake of simplifying the discussion, we will not address here, nor in Chapter 7, the memory constraints during process mapping and the implications of memory space in the incremental design process.

This chapter is organized as follows. The next section presents some issues related to mapping and scheduling in the context of a system based on a TDMA communication protocol. In Section 5.2 the problem we are going to solve is formulated. Section 5.3 introduces our approach to quantitatively characterize certain features of future applications. In Section 5.3 we introduce the metrics we have defined in order to capture the quality of a given design alternative and, based on these metrics, we give an exact problem formulation. Our mapping and scheduling strategy is described in Section 5.4 and the experimental results are presented in Section 5.5.

5.1 Background

In order to implement an application represented as a set of conditional process graphs as describe in Section 2.3, the designer has to map the processes to the system nodes and to derive a static cyclic schedule such that all deadlines are satisfied. We first illustrate some of the problems related to mapping and scheduling in the context of a system based on a TDMA communication protocol, before going on to explore further aspects specific to an incremental design approach.

Example 5.1: Let us consider the example in Figure 5.1 where we want to map an application consisting of four processes P_1 to P_4, with a period and deadline of 50 ms. The architecture is composed of three nodes that communicate according to a TDMA protocol, such that N_i transmits in slot S_i. For this example we suppose that there is no other previous application running on the system.

According to the specification, processes P_1 and P_3 are constrained to node N_1, while P_2 and P_4 can be mapped on nodes N_2 or N_3, but not N_1. The worst case execution times of processes on each potential node and the sequence and size of TDMA slots, are presented in Figure 5.1. In order to keep the example simple, we suppose that the message sizes are such that each message fits into one TDMA slot.

We consider two alternative mappings. If we map P_2 and P_4 on the faster processor N_3, the resulting schedule length (Figure 5.1a) will be 52 ms, which does not meet the deadline. However, if we map P_2 and P_4 on the slower processor N_2, the schedule length (Figure 5.1b) is 48 ms, which meets the deadline. Note, that the total traffic on the bus is the same for both mappings and the initial processor load is 0 on both N_2 and N_3.

This result has its explanation in the impact of the communication protocol. P_3 cannot start before receiving messages

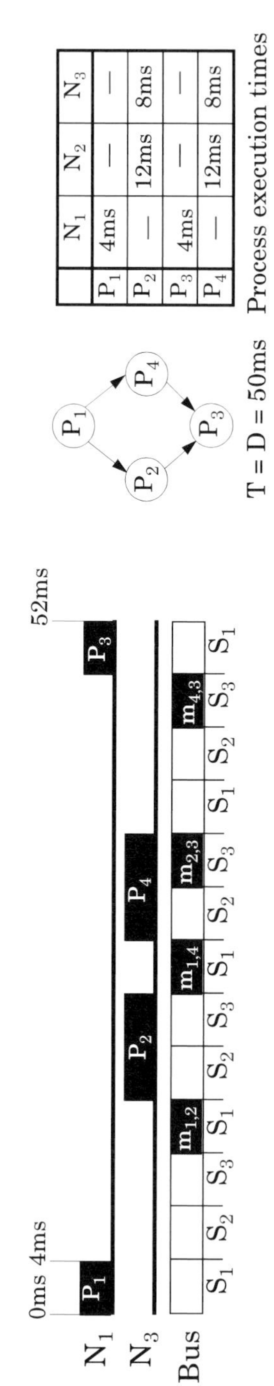

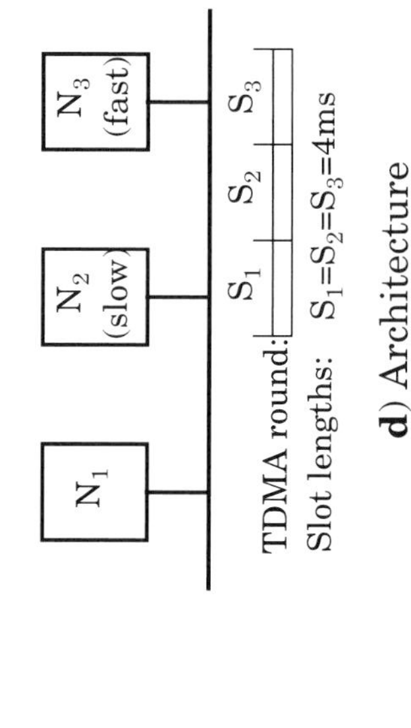

	N1	N2	N3
P1	4ms	—	—
P2	—	12ms	8ms
P3	4ms	—	—
P4	—	12ms	8ms

Figure 5.1: Mapping and Scheduling Examples for Time-Driven Systems

$m_{2,3}$ and $m_{4,3}$. However, slot S_2 corresponding to node N_2 precedes in the TDMA round slot S_3 on which node N_3 communicates. Thus, the messages which P_3 needs are available sooner in the case P_2 and P_4 are, counter-intuitively, mapped on the slower node.

■

But finding a valid schedule is not enough if we are to support an incremental design process as discussed in the introduction. In this case, starting from a valid design, we have to improve the mapping and scheduling so that not only the design constraints are satisfied, but also there is a good chance that, later, new functionality can easily be mapped on the resulted system.

Example 5.2: To illustrate the role of mapping and scheduling in the context of an incremental design process, let us consider the example in Figure 5.2. For simplicity, we consider an architecture consisting of a single processor. The system is currently running application ψ(Figure 5.2a).

At a particular moment application Γ_1 has to be implemented on top of ψ Three possible implementation alternatives for Γ_1 are depicted in Figure 5.2b$_1$, 5.2c$_1$, and 5.2d$_1$. All three are meeting the imposed time constraint for Γ_1.

At a later moment, application Γ_2 has to be implemented on the system running ψplus Γ_1. If Γ_1 has been implemented as shown in Figure 5.2b$_1$, there is no possibility to map application Γ_2 on the given system (in particular, there is no time slack available for process P_7). If Γ_1 has been implemented as in Figure 5.2c$_1$ or 5.2d$_1$, Γ_2 can be correctly mapped and scheduled on top of $\psi$and Γ_1.

■

There are two aspects which should be highlighted based on this example:

1. If application Γ_1 is implemented like in Figure 5.2c$_1$ or 5.2d$_1$, it is possible to implement Γ_2 on top of the existing system, without performing any modifications on the implementa-

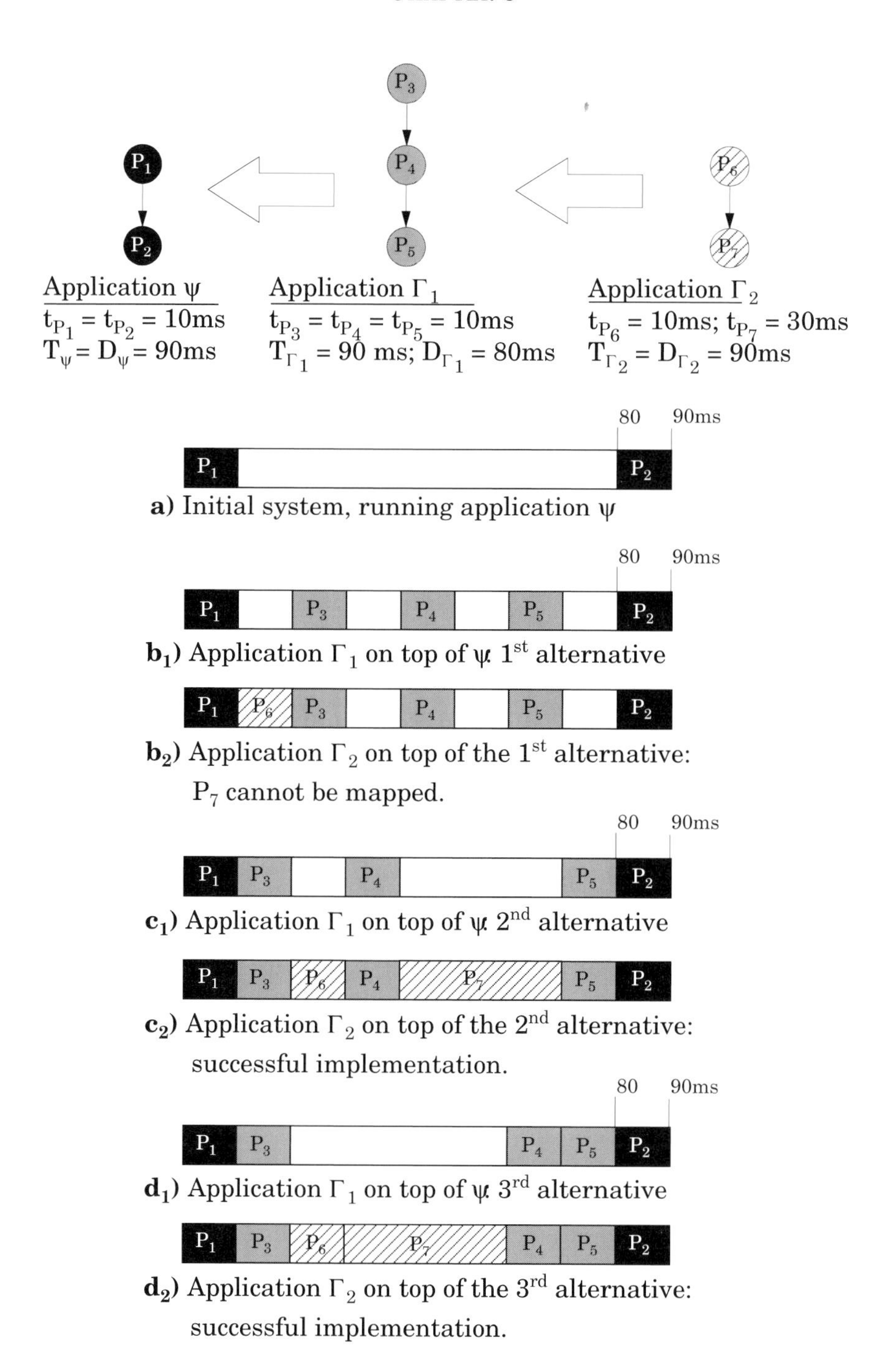

Figure 5.2: Incremental Mapping and Scheduling Examples for Time-Driven Systems

tion of previous applications. This could be the case if, during implementation of Γ_1, the designers have taken into consideration the fact that, in future, an application having the characteristics of Γ_2 will possibly be added to the system.

2. If Γ_1 has been implemented like in Figure 5.2b$_1$, Γ_2 can be added to the system only after performing certain modifications on the implementation of Γ_1 and/or ψ. In this case, of course, it is important to perform as few as possible modifications on previous applications, in order to reduce the development costs.

5.2 Incremental Mapping and Scheduling

Our goal is to map and schedule an application $\Gamma_{current}$ on a system that already implements a set ψ of applications, considering the following requirements:

Requirement *a*	All constraints on $\Gamma_{current}$ are satisfied and minimal modifications are performed to the applications in ψ
Requirement *b*	New applications Γ_{future} can be mapped on top of the resulting system.

In order to achieve our goal we need certain information to be available concerning the set of applications ψ as well as the possible future applications Γ_{future}. What exactly we have to know about existing applications has been outlined in Section 2.3.2, while the characterization of future applications will be discussed in the next section. In Section 5.3 we then introduce the quality metrics which will allow us to give a more rigorous formulation of the problem we are going to solve.

The processes in application $\Gamma_{current}$ can interact with the previously mapped applications ψ by reading messages generated on the bus by processes in ψ In this case, the reading process has to be synchronized with the arrival of the message on the bus, which is easy to model as an additional time constraint on the particular receiving process. This constraint is then considered (as any other deadline) during scheduling of $\Gamma_{current}$.

5.2.1 CHARACTERIZING FUTURE APPLICATIONS

What do we suppose to know about the family Γ_{future} of applications which do not exist yet? Given a certain limited application area (e.g., automotive electronics), it is not unreasonable to assume that, based on the designers' previous experience, the nature of expected future functions to be implemented, profiling of previous applications, available incomplete designs for future versions of the product, etc., it is possible to characterize the family of applications which possibly could be added to the current implementation. This is an assumption which is basic for the concept of incremental design.

Hence, we consider that, with respect to the future applications, we know the set $S_t = \{t_{min}, ..., t_i, ..., t_{max}\}$ of possible worst-case execution times for processes, and the set $S_b = \{b_{min}, ..., b_i, ..., b_{max}\}$ of possible message sizes. We also assume that over these sets we know the distributions of probability $f_{S_t}(t)$ for $t \in S_t$ and $f_{S_b}(b)$ for $b \in S_b$.

> **Example 5.3:** For example, we might have predicted possible worst-case execution times of different processes in future applications S_t={50, 100, 200, 300, 500 ms}. If there is a higher probability of having processes of 100 ms, and a very low probability of having processes of 300 ms and 500 ms, then our distribution function $f_{S_t}(t)$ could look like this: $f_{S_t}(50) = 0.20$, $f_{S_t}(100) = 0.50$, $f_{S_t}(200) = 0.20$, $f_{S_t}(300) = 0.05$, and $f_{S_t}(500) = 0.05$.
>
> ■

Another information concerning the future applications is related to the period of the constituent process graphs. In particular, the smallest expected period T_{min} is assumed to be given, together with the expected necessary processor time t_{need}, and bus bandwidth b_{need}, inside such a period T_{min}. As will be shown later, this information is treated in a flexible way during the design process and is used in order to provide a fair distribution of available resources.

The execution times in S_t, as well as t_{need}, are considered relative to the slowest node in the system. All the other nodes are characterized by a speedup factor relative to this slowest node. A normalization with these factors is performed when computing the metrics C_1^P and C_2^P introduced in the following section.

5.3 Quality Metrics and Objective Function

A designer will be able to map and schedule an application Γ_{future} on top of a system implementing ψ and $\Gamma_{current}$ only if there are sufficient resources available. For the discussion in this chapter, the resources which we consider are processor time and the bandwidth on the bus. In the context of a non-preemptive static scheduling policy, having free resources translates into having free time slots on the processors and having space left for messages in the bus slots. We call these free slots of available time on the processor or on the bus, *slack*.

It is to be noted that the total quantity of computation and communication power available on our system after we have mapped and scheduled $\Gamma_{current}$ on top of ψ is the same regardless of the mapping and scheduling policies used. What depends on the mapping and scheduling strategy is the distribution of slacks along the time line and the size of the individual slacks. It is exactly this size and distribution of the slacks that characterizes the quality of a certain design alternative from the point of view of flexibility for future upgrades.

In this section we introduce two criteria in order to reflect the degree to which a design alternative meets the requirement *b* presented in Section 5.2. For each criterion we provide metrics which quantify the degree to which the criterion is met. The first criterion reflects how well the resulted slack sizes fit to a future application, and the second criterion expresses how well the slack is distributed in time.

5.3.1 SLACK SIZES (THE FIRST CRITERION)

The slack sizes resulted after the implementation of $\Gamma_{current}$ on top of ψ should be such that they best accommodate a given family of applications Γ_{future}, characterized by the sets S_t, S_b and the probability distributions f_{S_t} and f_{S_b}, as outlined in Section 5.2.1.

Example 5.4: Let us go back to the example in Figure 5.2 where Γ_1 is what we now call $\Gamma_{current}$, while Γ_2, to be later implemented on top of ψ and Γ_1, is Γ_{future}. This Γ_{future} consists of the two processes P_6 and P_7. It can be observed that the best configuration from the point of view of accommodating Γ_{future}, taking into consideration only slack sizes, is to have a contiguous slack after implementation of $\Gamma_{current}$ (Figure 5.2d$_1$). However, in reality, it is almost impossible to map and schedule the current application such that a contiguous slack is obtained. Not only is it impossible, but it is also undesirable from the point of view of the second design criterion, to be discussed next. However, as we can see from Figure 5.2b$_1$, if we schedule $\Gamma_{current}$ such that it fragments too much the slack, it is impossible to fit Γ_{future} because there is no slack that can accommodate process P_7. A situation as the one depicted in Figure 5.2c$_1$ is desirable, where the resulted slack sizes are adapted to the characteristics of the Γ_{future} application.

■

In order to measure the degree to which the slack sizes in a given design alternative fit the future applications, we provide two metrics, C_1^P and C_1^m. C_1^P captures how much of the largest future application, which theoretically could be mapped on the system, can be mapped on top of the current design alternative. C_1^m is similar, relative to the slacks in the bus slots.

How does the largest future application which theoretically could be mapped on the system look like? The total processor time and bus bandwidth available for this largest future application is the total slack available on the processors and bus, respectively, after implementing $\Gamma_{current}$. Process and message sizes of this hypothetical largest application are estimated knowing the total size of the available slack, and the characteristics of the future applications as expressed by the sets S_t and S_b, and the probability distributions f_{S_t} and f_{S_b}.

Example 5.5: Let us consider, for example, that the total slack size on the processors is of 2800 ms and the set of possible worst case execution times is S_t = {50, 100, 200, 300, 500 ms}. The probability distribution function f_{S_t} is defined as follows: $f_{S_t}(50) = 0.20$, $f_{S_t}(100) = 0.50$, $f_{S_t}(200) = 0.20$, $f_{S_t}(300) = 0.05$, and $f_{S_t}(500) = 0.05$. Under these circumstances, the largest hypothetical future application will consist of 20 processes: 10 processes (half of the total, $f_{S_t}(100) = 0.50$) with a worst case execution time of 100 ms, four processes with 50 ms, four with 200 ms, one with 300 and one with 500 ms.

■

After we have determined the number of processes of this largest hypothetical Γ_{future} and their worst-case execution times, we apply a *bin-packing* algorithm [Mar90] using the *best-fit* policy in which we consider processes as the objects to be packed, and the available slacks as containers. The total execution time of processes which are left unpacked, relative to the total execution time of the whole process set, gives the metric C_1^P. The same is the case with the metric C_1^m, but applied to message sizes and available slacks in the bus slots.

Example 5.6: Let us consider the example in Figure 5.2 and suppose a hypothetical Γ_{future} consisting of two processes like those of application Γ_2. For the design alternatives in Figure 5.2c$_1$ and 5.2d$_1$, $C_1^P = 0\%$ (both alternatives are perfect from the point of view of slack sizes). For the alternative in Figure 5.2b$_1$, however, C_1^P= 30 / 40 = 75% — the worst case execution time of P_7 (which is left unpacked) relative the total execution time of the two processes.

■

5.3.2 DISTRIBUTION OF SLACKS (THE SECOND CRITERION)

In the previous section we have defined a metric which captures how well the sizes of the slacks fit a possible future application. A similar metric is needed to characterize the distribution of slacks over time.

Let P_i be a process with period T_i that belongs to a future application, and $M(P_i)$ the node on which P_i will be mapped. The worst case execution time of P_i on node $M(P_i)$ is C_i. In order to schedule P_i we need a slack of size C_i that is available periodically, within a period T_i, on processor $M(P_i)$. If we consider a group of processes with period T, which are part of Γ_{future}, in order to implement them, a certain amount of slack is needed which is available periodically, with a period T, on the nodes implementing the respective processes.

During the implementation of $\Gamma_{current}$ we aim for a slack distribution such that the future application with the smallest expected period T_{min} and with the necessary processor time t_{need}, and bus bandwidth b_{need}, can be accommodated (see Section 5.2.1).

Thus, for each node, we compute the minimum periodic slack, inside a T_{min} period. By summing these minima, we obtain the slack which is available periodically to Γ_{future}. This is the C_2^P metric. The C_2^m metric characterizes the minimum periodically available bandwidth on the bus and it is computed in a similar way.

Example 5.7: In Figure 5.3 we consider an example with T_{min} = 120 ms, t_{need} = 90 ms, and b_{need} = 65 ms. The length of the schedule table of the system implementing ψ and $\Gamma_{current}$ is 360 ms (in Section 5.4 we will elaborate on the length of

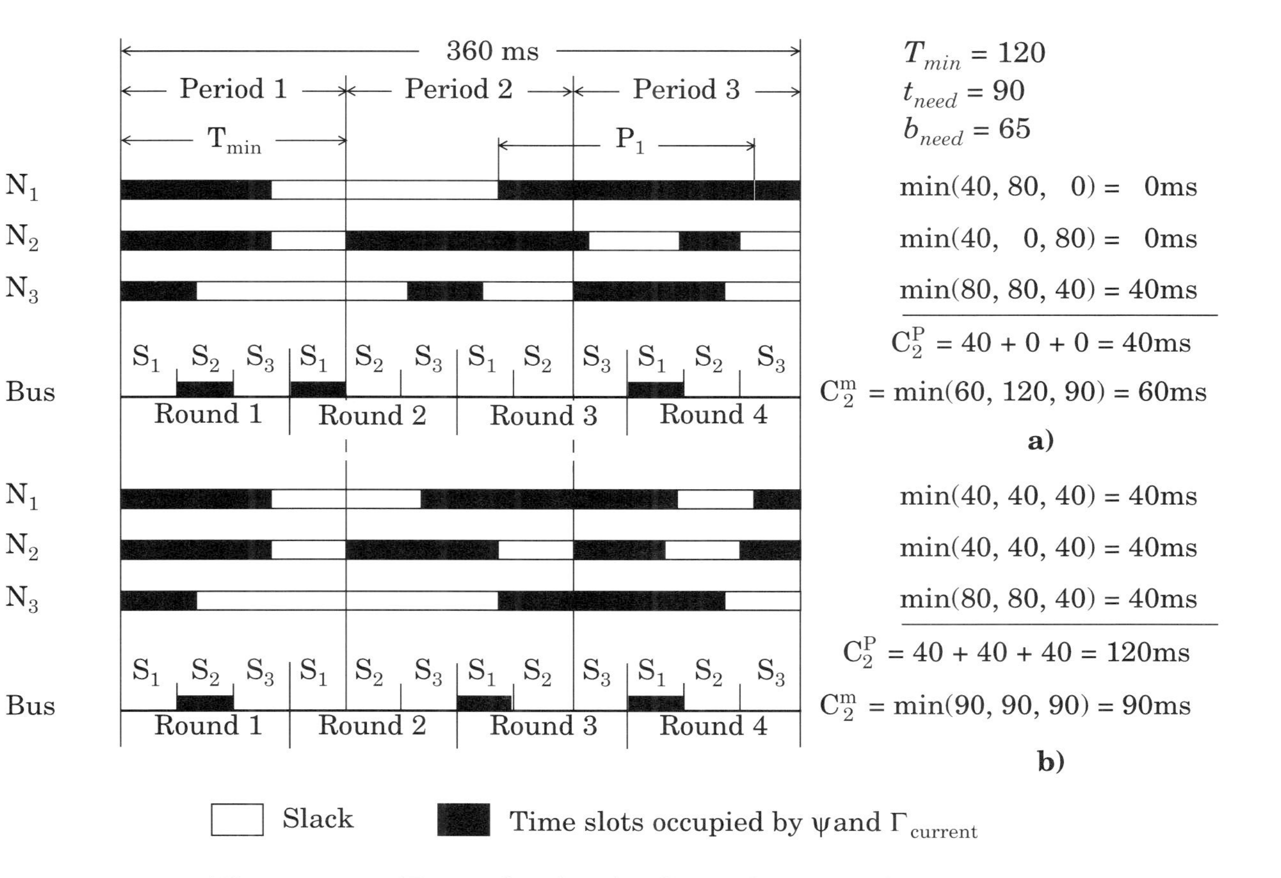

Figure 5.3: Examples for the Second Design Criterion

the global schedule table). Consequently, we have to investigate three periods of length T_{min} each. The system consists of three nodes.

Let us consider the situation in Figure 5.3a. In the first period, *Period 1*, there are 40 ms of slack available on node N_1, in the second period 80 ms, and in the third period no slack is available on N_1. Hence, the total slack a future application of period T_{min} can use on node N_1 is min(40, 80, 0) = 0 ms. Neither can node N_2 provide slack for this application, as in *Period 1* there is no slack available. However, on node N_3 there are at least 40 ms of slack available in each period. Thus, with the configuration in Figure 5.3a we have $C_2^P = 40$ ms, which is not sufficient to accommodate $t_{need} = 90$ ms. The available periodic slack on the bus is also insufficient: $C_2^m = 60 \text{ ms} < b_{need}$.

However, in the situation presented in Figure 5.3b, we have $C_2^P = 120 \text{ ms} > t_{need}$, and $C_2^m = 90 \text{ ms} > b_{need}$, which means that enough resources are available, periodically, for the application.

■

5.3.3 Objective Function and Exact Problem Formulation

In order to capture how well a certain design alternative meets the requirement *b* stated in Section 5.2, the metrics discussed before are combined in an objective function, as follows:

$$C = w_1^P (C_1^P)^2 + w_1^m (C_1^m)^2 + w_2^P max(0, t_{need} - C_2^P) + w_2^m max(0, b_{need} - C_2^m) \quad (5.1)$$

where the metric values introduced in the previous section are weighted by the constants w_1^P, w_2^P, w_1^m, and w_2^m. Our mapping and scheduling strategy will try to minimize this function.

The first two terms measure how well the resulted slack sizes fit to a future application (the first criterion), while the second two terms reflect the distribution of slacks (the second criterion). In order to obtain a balanced solution, that favors a good fitting both on the processors and on the bus, we have used the squares of the metrics.

We call a *valid solution* one with a mapping and scheduling which satisfies all the design constraints (in our case the deadlines) and meets the second criterion ($C_2^P \geq t_{need}$ and $C_2^m \geq b_{need}$)[1].

At this point we can give an exact formulation of our problem: Given an existing set of applications ψ which are already mapped and scheduled, and an application $\Gamma_{current}$ to be implemented on top of ψ we are interested to find that subset $\Omega \subseteq \psi$ of old applications to be remapped and rescheduled such that we produce a valid solution for $\Gamma_{current} \cup \Omega$ and the total cost of modification $R(\Omega)$ is minimized (see Section 2.3.2 for the details concerning the modification cost of an application). Once such a set Ω of applications is found, we are interested to optimize the implementation of $\Gamma_{current} \cup \Omega$ such that the objective function C (Equation 5.1) is minimized, considering a family of future applications characterized by the sets S_t and S_b, the functions f_{S_t} and f_{S_b} as well as the parameters T_{min}, t_{need}, and b_{need}.

A mapping and scheduling strategy based on this problem formulation is presented in the following section.

5.4 Mapping and Scheduling Strategy

As shown in the algorithm in Figure 5.4, our mapping and scheduling strategy (MS) consists of two steps. In the first step (lines 1–14) we try to obtain a valid solution for the mapping and scheduling of $\Gamma_{current} \cup \Omega$ so that the modification cost $R(\Omega)$ is

1. This definition of a valid solution can be relaxed by imposing only the satisfaction of deadlines. In this case, the mapping and scheduling algorithm in Figure 5.4 will look after a solution which satisfies the deadlines and minimizes $R(\Omega)$; the additional second criterion is, in this case, only considered optionally.

minimized. Starting from such a solution, the second step (lines 17–20) iteratively improves the design in order to minimize the objective function C. In the context in which the second criterion is satisfied after the first step, improving the cost function during the second step aims at minimizing the value of $w_1^P (C_1^P)^2 + w_1^m (C_1^m)^2$.

If the first step has not succeeded in finding a solution such that the imposed timing constraints are satisfied, this means that there are not sufficient resources available to implement the application $\Gamma_{current}$. Thus, modifications of the system architecture have to be performed before restarting the mapping and scheduling procedure. If, however, the timing constraints are met but the second design criterion is not satisfied, a larger T_{min}

MappingSchedulingStrategy

1 **Step 1**: try to find a valid solution that minimizes $R(\Omega)$
2   Find a mapping and scheduling of $\Gamma_{current} \cup \Omega$ on top of $\psi \setminus \Omega$ so that:
3   1. constraints are satisfied;
4   2. modification cost $R(\Omega)$ is minimized;
5   3. the second criterion is satisfied: $C_2^P \geq t_{need}$ and $C_2^m \geq b_{need}$
6
7 **if** Step1 has not succeeded **then**
8   **if** constraints are not satisfied **then**
9   change architecture
10   **else**
11   suggest new T_{min}, t_{need} or b_{need}
12   **end if**
13   **go to** Step 1
14 **end if**
15
16
17 **Step 2**: improve the solution by minimizing objective function C
18   Perform iteratively transformations which
19   improve the first criterion (the metrics C_1^P and C_1^m)
20   without invalidating the second criterion.
21
end MappingSchedulingStrategy

Figure 5.4: The Mapping and Scheduling Strategy

(smallest expected period of a future application, see Section 5.2.1) or smaller values for t_{need} and/or b_{need} are suggested to the designer (line 11). This, of course, reduces the frequency of possible future applications and the amount of processor and bus resources available to them.

In the following section we briefly discuss the basic mapping and scheduling algorithm we have used in order to generate an initial solution. The heuristic used to iteratively improve the design with regard to the first and the second design criteria is presented in Section 5.4.2. In Section 5.4.3 we describe three alternative heuristics which can be used during the first step in order to find the optimal subset of applications to be modified.

5.4.1 THE INITIAL MAPPING AND SCHEDULING

The first step of our mapping and scheduling strategy MS consists of an iteration that tries different subsets $\Omega \subseteq \psi$ with the intention to find that subset $\Omega = \Omega_{min}$ of old applications to be remapped and rescheduled which produces a valid solution for $\Gamma_{current} \cup \Omega$ such that $R(\Omega)$ is minimized. Given a subset Ω, the InitialMappingScheduling function (IMS) constructs a mapping and a schedule for the applications $\Gamma_{current} \cup \Omega$ on top of $\psi \setminus \Omega$, which meets the deadlines, without worrying about the two criteria introduced in Section 5.3.

The IMS is a classical mapping and scheduling algorithm for which we have used as a starting point the Heterogeneous Critical Path (HCP) algorithm, introduced in [Jor97]. The HCP is based on a list scheduling approach [Cof72]. We have modified the HCP algorithm in four main regards:

1. The list scheduling approach that is used as a basis for HCP is considering applications modeled as conditional process graphs, as described in Section 4.2.1.
2. We consider that mapping and scheduling does not start with an empty system but a system on which a certain number of processes are already mapped.

3. Messages are scheduled into bus-slots according to the TDMA protocol. The TDMA-based message scheduling technique has been presented in Section 4.3.
4. As a priority function for list scheduling we use, instead of the CP (critical path) priority function employed in [Jor97], the MPCP (modified partial critical path) function introduced in Section 4.3.2. The MPCP takes into consideration the particularities of the communication protocol for calculation of communication delays. These delays are not estimated based only on the message length, but also on the time when slots, assigned to the particular node which generates the message, will be available.

For the example in Figure 5.1, our initial mapping and scheduling algorithm will be able to produce the optimal solution with a schedule length of 48 ms.

However, before performing the effective mapping and scheduling with IMS, two aspects have to be addressed. First, the process graphs $G_i \in \Gamma_{current} \cup \Omega$ have to be merged into a single graph $G_{current}$, by unrolling of process graphs and inserting dummy nodes as shown in Figure 5.5. The period $T_{G_{current}}$ of $G_{current}$ is equal to the least common multiplier of the periods T_{G_i} of the graphs G_i. Dummy nodes (depicted as empty disks in Figure 5.5) represent processes with a certain execution time but are not mapped to any processor or bus.

In addition, we have to consider during scheduling the mismatch between the periods of the already existing system and those of the current application. The schedule table into which we would like to schedule $G_{current}$ has a length of $T_{\psi\backslash\Omega}$ which is the global period of the system ψ after extraction of the applications in Ω. However, the period $T_{current}$ of $G_{current}$ can be different from $T_{\psi\backslash\Omega}$. Hence, before scheduling $G_{current}$ into the existing schedule table, the schedule table is expanded to the least common multiplier of the two periods. A similar procedure is followed in the case $T_{current} > T_{\psi\backslash\Omega}$.

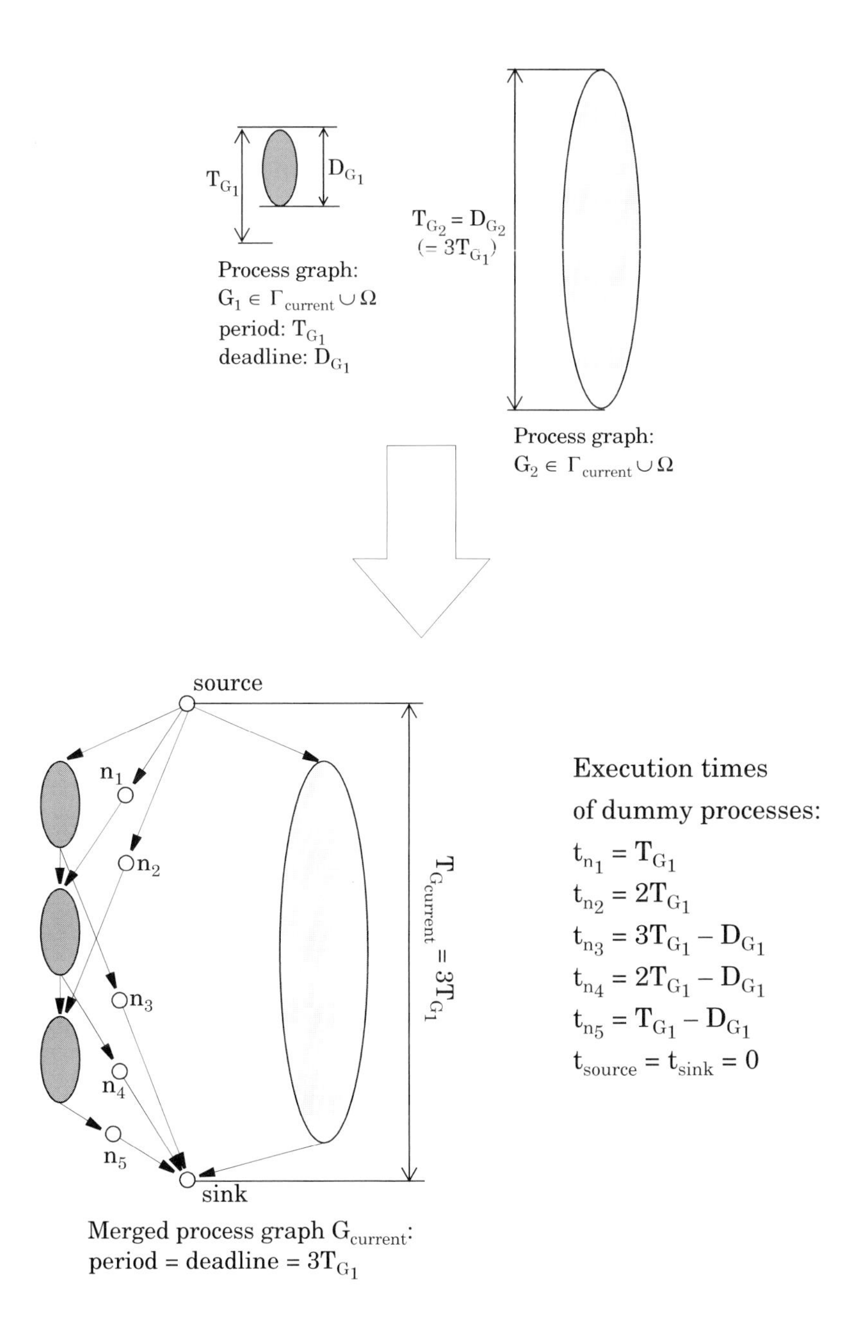

Figure 5.5: Process Graph Merging Example

5.4.2 ITERATIVE DESIGN TRANSFORMATIONS

Once IMS has produced a mapping and scheduling which satisfies the timing constraints, the next goal of Step 1 is to improve the design in order to satisfy the second design criterion ($C_2^P \geq t_{need}$ and $C_2^m \geq b_{need}$). During the second step, the design is then further transformed with the goal of minimizing the value of $w_1^P(C_1^P)^2 + w_1^m(C_1^m)^2$, according to the requirements of the first criterion, without invalidating the second criterion achieved in the first step. In both steps we iteratively improve the design using a transformational approach. These successive transformations are performed inside the (innermost) repeat loops of the first (lines 11–19 in Figure 5.6) and second step (lines 31–38). A new design is obtained from the current one by performing a transformation called *move*. We consider the following two categories of moves:

1. moving a process to a different slack found on the same node or on a different node;
2. moving a message to a different slack on the bus.

In order to eliminate those moves that will lead to an infeasible design (that violates deadlines), we do as follows. For each process P_i, we calculate the $ASAP(P_i)$ and $ALAP(P_i)$ times considering the resources of the given hardware architecture. $ASAP(P_i)$ is the earliest time P_i can start its execution, while $ALAP(P_i)$ is the latest time P_i can start its execution without causing the application to miss its deadline. When moving P_i we will consider slacks on the target processor only inside the [$ASAP(P_i)$, $ALAP(P_i)$] interval. The same reasoning holds for messages, with the addition that a message can only be moved to slacks belonging to a slot that corresponds to the sender node. Any violation of the data dependency constraints caused by a move is rectified by shifting processes or messages concerned in an appropriate way. If such a shift produces a deadline violation, the move is rejected.

```
1   Step 1: try to find a valid solution that minimizes R(Ω)
2       Ω=∅
3       repeat
4           succeeded=InitialMappingScheduling(ψ\ Ω, Γ_current ∪ Ω)
5           -- compute ASAP–ALAP intervals for all processes
6           ASAP(Γ_current ∪ Ω); ALAP(Γ_current ∪ Ω)
7           -- if time constraints are satisfied
8           if succeeded then
9               -- design transformations in order to satisfy
10              -- the second design criterion
11              repeat
12                  -- find set of moves with the highest potential
13                  -- to maximize C_2^P or C_2^m
14                  move_set = PotentialMoveC_2^P(Γ_current ∪ Ω) ∪
15                      PotentialMoveC_2^m(Γ_current ∪ Ω)
16                  -- select and perform move which improves most C_2
17                  move = SelectMoveC_2(move_set); Perform(move)
18                  succeeded = C_2^P ≥ t_need and C_2^m ≥ b_need
19              until succeeded or maximum number of iterations reached
20          end if
21          if succeeded and R(Ω) smallest so far then
22              Ω_valid = Ω, solution_valid = solution_current
23          end if
24          -- try another subset
25          Ω = NextSubset(Ω)
26      until termination condition
27
28  Step 2: improve the solution by minimizing objective function C
29      solution_current = solution_valid; Ω_min = Ω_valid
30      -- design transformations in order to satisfy the first design criterion
31      repeat
32          -- find set of moves with highest potential to minimize C_1^P or C_1^m
33          move_set = PotentialMoveC_1^P(Γ_current ∪ Ω_min) ∪
34              PotentialMoveC_1^m(Γ_current ∪ Ω_min)
35          -- select move which improves w_1^P(C_1^P)^2 + w_1^m(C_1^m)^2 and
36          -- does not invalidate the second criterion
37          move = SelectMoveC_1(move_set); Perform(move)
38      until w_1^P(C_1^P)^2 + w_1^m(C_1^m)^2 has not changed or
39          maximum number of iterations reached
```

Figure 5.6: Step One and Two of the Mapping and Scheduling Strategy in Figure 5.4

At each step, our heuristic tries to find those moves that have the highest potential to improve the design. For each iteration a set of potential moves is selected by the PotentialMoveX functions. SelectMoveX then evaluates these moves with regard to the respective metrics and selects the best one to be performed. We now briefly discuss the four PotentialMoveX functions with the corresponding moves.

PotentialMoveC_2^P *and PotentialMove*C_2^m

Example 5.8: Consider Figure 5.3a on page 105. In *Period 3* on node N_1 there is no available slack. However, if we move process P_1 with 40 ms to the left into *Period 2*, as depicted in Figure 5.3b, we create a slack in *Period 3* and the periodic slack on node N_1 will be min(40, 40, 40) = 40 ms, instead of 0 ms.
■

Potential moves aimed at improving the metric C_2^P will be the shifting of processes inside their [*ASAP, ALAP*] interval in order to improve the periodic slack. The move can be performed on the same node or to the less loaded nodes. The same is true for moving messages in order to improve the metric C_2^m. For the improvement of the periodic bandwidth on the bus, we also consider movement of processes, trying to place the sender and receiver of a message on the same processor and, thus, reducing the bus load.

PotentialMoveC_1^P *and PotentialMove*C_1^m

The moves suggested by these two functions aim at improving the C_1 metric through reducing the slack fragmentation. The heuristic is to evaluate only those moves that iteratively eliminate the smallest slack in the schedule.

Example 5.9: Let us consider the example in Figure 5.7, where we have three applications mapped on a single processor: ψ consisting of P_1 and P_2, $\Gamma_{current}$, having processes P_3, P_4 and P_5, and Γ_{future}, with P_6, P_7 and P_8. Figure 5.7 presents three possible schedules; processes are depicted with rectangles, the width of a rectangle representing the worst case execution time of that process. The PotentialMoveC_1 functions start by identifying the smallest slack in the schedule table.

In Figure 5.7a, the smallest slack is the slack between P_1 and P_3. Once the smallest slack has been identified, potential moves are investigated which either remove or enlarge the slack. For example, the slack between P_1 and P_3 can be removed by attaching P_3 to P_1, and it can be enlarged by moving P_3 to the right in the schedule table. Moves that remove the slack are considered only if they do not lead to an invalidation of the second design criterion, measured by the C_2 metric improved in the previous step (see Figure 5.6, Step 1). Also, the slack can be enlarged only if it does not create, as a result, other unusable slack. A slack is unusable if it

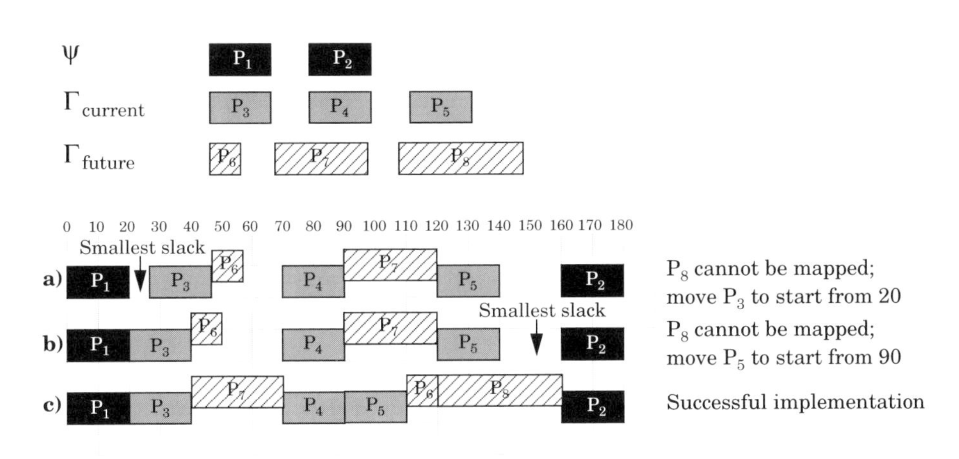

a)
Smallest slack: between P_1 and P_3
Potential moves: P_3 starting at 20, having C_1^P = 50% (denoted with 20/50%), 30/50%, 40/50%, 50/50%.
Selected move: P_3 to 20, with C_1^P= 50%.

b)
Smallest slack: between P_5 and P_2
Potential moves: P_5 to 90/0%, 100/0%, 110/50%, 130/50%, 140/50%, 150/0%, 160/0%.
Selected move: P_5 to 90 with C_1^P= 100%.

Figure 5.7: Successive Steps with Potential Moves for Improving the First Design Metric

cannot hold the smallest object of the future application, in our case P_6.

In Figure 5.7a, the slack can be removed by moving P_3 such that it starts from time 20, immediately after P_1, and it can be enlarged by moving P_3 so that it starts from 30, 40, or 50 (considering an increment which here was set by us to 10, the size of P_6, the smallest object in Γ_{future}). For each move, the improvement on the C_1 metric is calculated, and that move is selected by the SelectMoveC$_1$ function to be performed, which leads to the largest improvement on C_1. For all the previously considered moves of P_3, we are not able to map P_8 which represents 50% of the Γ_{future}, therefore C_1 = 50%. Consequently, we can perform any of the mentioned moves, and our algorithm selects the first one investigated, the move to start P_2 from 20, thus removing the slack. As a result of this move, the new schedule table is the one in Figure 5.7b.

In the next call to the PotentialMoveC$_1$ function, the slack between P_5 and P_2 is identified as the smallest slack. Out of the potential moves that eliminate this slack, listed in Figure 5.7 for case b, several lead to C_1 = 0%, the largest improvement (no processes from Γ_{future} are left out, so C_1 = 0%). SelectMoveC$_1$ selects moving P_5 to start from 90, and thus we are able to map process P_8 of the future application, leading to a successful implementation in Figure 5.7c.

■

The previous example has only illustrated movements of processes. Similarly, in PotentialMoveC$_1^m$, we also consider moves of messages in order to improve C_1^m. However, the movement of messages is restricted by the TDMA bus access scheme, such that a message can only be moved in the same slot of another round.

5.4.3 MINIMIZING THE TOTAL MODIFICATION COST

The first step of our mapping and scheduling strategy, described in Figure 5.6, iterates on successive subsets Ω searching for a valid solution, which also minimizes the total modification cost $R(\Omega)$ calculated using the Equation 2.1 in Section 2.3.2. As a first attempt, the algorithm searches for a valid implementation of $\Gamma_{current}$ without disturbing the existing applications ($\Omega = \emptyset$). If no valid solution is found, successive subsets Ω produced by the function NextSubset are considered, until a termination condition is met. The performance of the algorithm, in terms of runtime and quality of the solutions produced, is strongly influenced by the strategy employed for the function NextSubset and the termination condition. They determine how the design space is explored while testing different subsets Ω of applications.

In the following sections we present three alternative strategies for the implementation of the NextSubset function. The first two can be considered as situated at opposite extremes: The first one is potentially very slow but produces the optimal result while the second is very fast and possibly low quality. The third alternative is a heuristic capable of producing good quality results in relatively short time, as will be demonstrated by the experimental results presented in Section 5.5.2.

Exhaustive Search (ES)

In order to find Ω_{min}, the simplest solution is to try successively all the possible subsets $\Omega \subseteq \psi$ These subsets are generated in the ascending order of the total modification cost, starting from $\emptyset$. The termination condition is fulfilled when the first valid solution is found or no new subsets are to be generated. Since the subsets are generated in ascending order, according to their cost, the subset Ω that first produces a valid solution is also the subset with the minimum modification cost.

The generation of subsets is performed according to the graph $\mathcal{A}$ that characterizes the existing applications (see Section 2.3.2). Finding the next subset Ω, starting from the current one, is achieved by a branch and bound algorithm that, in the worst case, grows exponentially in time with the number of applications.

Example 5.10: For the example in Figure 2.6 on page 34, discussed in Section 2.3.2, the call to NextSubset(∅) will generate $\Omega = \{\Gamma_7\}$ which has the smallest nonzero modification cost $R(\{\Gamma_7\}) = 20$. The next generated subsets, in order, together with their corresponding total modification costs are: $R(\{\Gamma_3\}) = 50$, $R(\{\Gamma_3, \Gamma_7\}) = 70$, $R(\{\Gamma_4, \Gamma_7\}) = 90$ (the inclusion of Γ_4 triggers the inclusion of Γ_7), $R(\{\Gamma_2, \Gamma_3\}) = 120$, $R(\{\Gamma_2, \Gamma_3, \Gamma_7\}) = 140$, $R(\{\Gamma_3, \Gamma_4, \Gamma_7\}) = 140$, $R(\{\Gamma_1\}) = 150$, and so on. The total number of possible subsets according to the graph $\mathcal{A}$ in Figure 2.6 is 16.

■

This approach, while finding the optimal subset Ω, requires a large amount of computation time and can be used only with a small number of applications.

Ad-Hoc Selection Heuristic (AS)

If the number of applications is larger, a possible solution could be based on a simple greedy heuristic which, starting from $\Omega = \varnothing$, progressively enlarges the subset until a valid solution is produced. The algorithm looks at all the non-frozen applications and picks that one which, together with its dependencies, has the smallest modification cost. If the new subset does not produce a valid solution, it is enlarged by including, in the same fashion, the next application with its dependencies. This greedy expansion of the subset is continued until the set is large enough to lead to a valid solution or no application is left.

Example 5.11: For the example in Figure 2.6 the call to NextSubset(∅) will produce $R(\{\Gamma_7\}) = 20$, and will be successively enlarged to $R(\{\Gamma_7, \Gamma_3\}) = 70$, $R(\{\Gamma_7, \Gamma_3, \Gamma_2\}) = 140$ (Γ_4 could have been picked as well in this step because it has the same modification cost of 70 as Γ_2 and its dependency Γ_7 is already in the subset), $R(\{\Gamma_7, \Gamma_3, \Gamma_2, \Gamma_4\}) = 210$, and so on. ■

While this approach finds very quickly a valid solution, if one exists, it is possible that the resulted total modification cost is much higher than the optimal one.

Subset Selection Heuristic (SH)

An intelligent selection heuristic should be able to identify the reasons due to which a valid solution has not been produced in the first step of the MS algorithm in Figure 5.6, and to find the set of candidate applications which, if modified, could eliminate the problem.

The failure to produce a valid solution can have two possible causes: an initial mapping which meets the deadlines has not been found, or the second criterion is not satisfied.

Let us investigate the first reason. If an application Γ_i is to meet its deadline D_i, all its processes $P_j \in \Gamma_i$ have to be scheduled inside their [*ASAP*, *ALAP*] intervals. InitialMappingScheduling (IMS) fails to schedule a process inside its [*ASAP*, *ALAP*] interval, if there is not enough slack available on any processor, due to other processes scheduled in the same interval. In this situation we say that there is a *conflict* with processes belonging to other applications. We are interested to find out which applications are responsible for conflicts encountered during the mapping and scheduling of $\Gamma_{current}$, and not only that, but also which ones are *flexible* enough to be moved away in order to avoid these conflicts.

If it is not able to find a solution that satisfies the deadlines, IMS will determine a metric Δ_{Γ_i} that characterizes both the degree of conflict and the flexibility of each application $\Gamma_i \in \psi$ in relation to $\Gamma_{current}$. A set of applications Ω will be characterized, in relation to $\Gamma_{current}$, by the following metric:

$$\Delta(\Omega) = \sum_{\Gamma_i \in \Omega} \Delta_{\Gamma_i}. \quad (5.2)$$

This metric $\Delta(\Omega)$ will be used by our subset selection heuristic in the case IMS has failed to produce a solution which satisfies the deadlines. An application with a larger Δ_{Γ_i} is more likely to lead to a valid schedule if included in Ω.

Example 5.12: In Figure 5.8 we illustrate how this metric is calculated. Applications A, B and C are implemented on a system consisting of the three processors N_1, N_2 and N_3. The current application to be implemented is D. At a certain moment, IMS comes to the point to map and schedule process $D_1 \in D$. However, it is not able to place it inside its [*ASAP*, *ALAP*] interval, denoted in Figure 5.8 as I. The reason is that there is not enough slack available inside I on any of the processors, because processes $A_1, A_2, A_3 \in A$, $B_1 \in B$, and $C_1 \in C$ are scheduled inside that interval. We are interested to determine which of the applications A, B, and C are more likely to lend free slack for D_1, if remapped and rescheduled.

Therefore, we calculate the slack resulted after we move away processes belonging to these applications from the interval I. For example, the resulted slack available after modifying application C (moving C_1 either to the left or to the right inside its own [*ASAP*, *ALAP*] interval) is of size $|I| - \min(|C_1^L|, |C_1^R|)$. With C_1^L (C_1^R) we denote that slice of process C_1 which remains inside the interval I after C_1 has been moved to the extreme left (right) inside its own [*ASAP*, *ALAP*] interval. $|C_1^L|$ represents the length of slice C_1^L. Thus, when considering process D_1, Δ_C will be incremented with $\delta_C^{D_1} = \max(|I| - \min(|C_1^L|, |C_1^R|) - |D_1|, 0)$. This value shows the maximum theoretical slack usable for D_1, that can be produced by modifying the application C. By relating this slack to the length of D_1, the value $\delta_C^{D_1}$ also captures the amount of flexibility provided by that modification.

The increments $\delta_B^{D_1}$ and $\delta_A^{D_1}$ to be added to the values of Δ_B and Δ_A respectively, are also presented in Figure 5.8. IMS then continues the evaluation of the metrics Δ with the other processes belonging to the current application D (with the assumption that process D_1 has been scheduled at the beginning of interval I). Thus, as result of the failed attempt to map and schedule application D, the metrics Δ_A, Δ_B, and Δ_C will be produced.

■

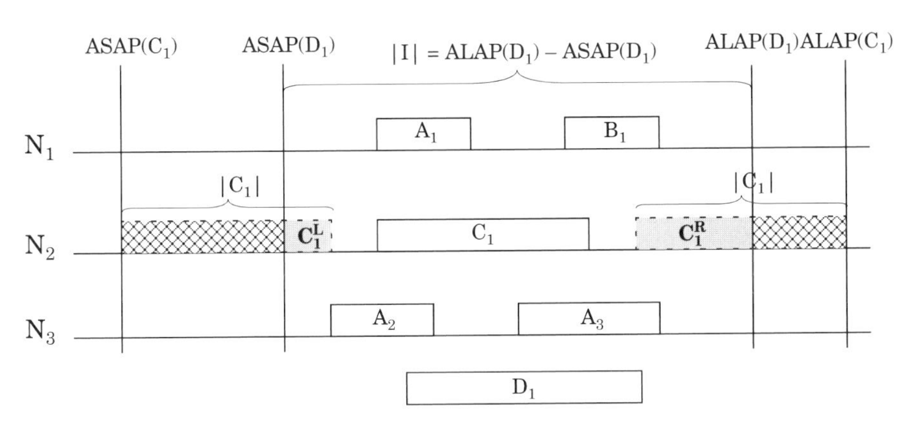

$$\delta_A^{D_1} = \max(\overbrace{\max(|I| - |B_1| - \min(|A_1^L|, |A_1^R|)}^{D_1 \text{ mapped on } N_1},\ \underbrace{|I| - \min(|A_2^L|, |A_2^R|) - \min(|A_3^L|, |A_3^R|)) \text{ P } |D_1|, 0)}_{D_1 \text{ mapped on } N_3}$$

$$\delta_B^{D_1} = \max(|I| - |A_1| - \min(|B_1^L|, |B_1^R|) - |D_1|, 0);$$

$$\delta_C^{D_1} = \max(|I| - \min(|C_1^L|, |C_1^R|) - |D_1|, 0)$$

Figure 5.8: Metric for the Subset Selection Heuristic

If the initial mapping was successful, the first step of MS could fail during the attempt to satisfy the second criterion (Figure 5.6). In this case, the metric Δ_{Γ_i} is computed in a different way. What Δ_{Γ_i} will capture in this case, is the potential of an application Γ_i to improve the metric C_2 if remapped together with $\Gamma_{current}$. Therefore, we consider a total number of moves from all the non-frozen applications in ψ These moves are determined using the PotentialMoveC$_2$ functions presented in Section 5.4.2. Each such move will lead to a different mapping and schedule, and thus to a different C_2 value. Let us consider δ_{move} as the improvement on C_2 produced by the currently considered move. If there is no improvement, $\delta_{move} = 0$. Thus, for each move that has as subject P_j or $m_j \in \Gamma_i$, we increment the metric Δ_{Γ_i} with the δ_{move} improvement on C_2.

As shown in the algorithm in Figure 5.6, MS starts by trying an implementation of $\Gamma_{current}$ with $\Omega = \emptyset$. If this attempt fails, because of one of the two reasons mentioned above, the corresponding metrics Δ_{Γ_i} are computed for all $\Gamma_i \in \psi$

Our heuristic SH will then start by finding the solution Ω_{AS} produced with the greedy heuristic AS (this will succeed if there exists any solution). The total modification cost corresponding to this solution is $R_{AS} = R(\Omega_{AS})$ and the value of the metric Δ is $\Delta_{AS} = \Delta(\Omega_{AS})$.

SH now continues by trying to find a solution with a more favorable Ω than Ω_{AS} (a smaller total cost R). Therefore, the thresholds $R_{max} = R_{AS}$ and $\Delta_{min} = \Delta_{AS} / n$ (for our experiments we considered $n = 2$) are set. Sets of applications not fulfilling these thresholds will not be investigated by MS.

For generating new subsets Ω, the function NextSubset now follows a similar approach like in the exhaustive search approach ES, but in a reverse direction, towards smaller subsets (starting with the set containing all non-frozen applications), and it will consider only subsets with a smaller total cost than R_{max} and a larger Δ than Δ_{min} (a small Δ means a reduced potential to eliminate the cause of the initial failure).

Each time a valid solution is found, the current values of R_{max} and Δ_{min} are updated in order to further restrict the search space. The heuristic stops when no subset can be found with $\Delta > \Delta_{min}$, or a certain imposed limit has been reached (e.g., on the total number of attempts to find new subsets).

5.5 Experimental Evaluation

In the following three sections we show a series of experiments that demonstrate the effectiveness of the proposed approaches and algorithms. The first set of results is related to the efficiency of our mapping and scheduling algorithm and the iterative design transformations proposed in Sections 5.4.1 and 5.4.2. The second set of experiments evaluates our heuristics for minimiza-

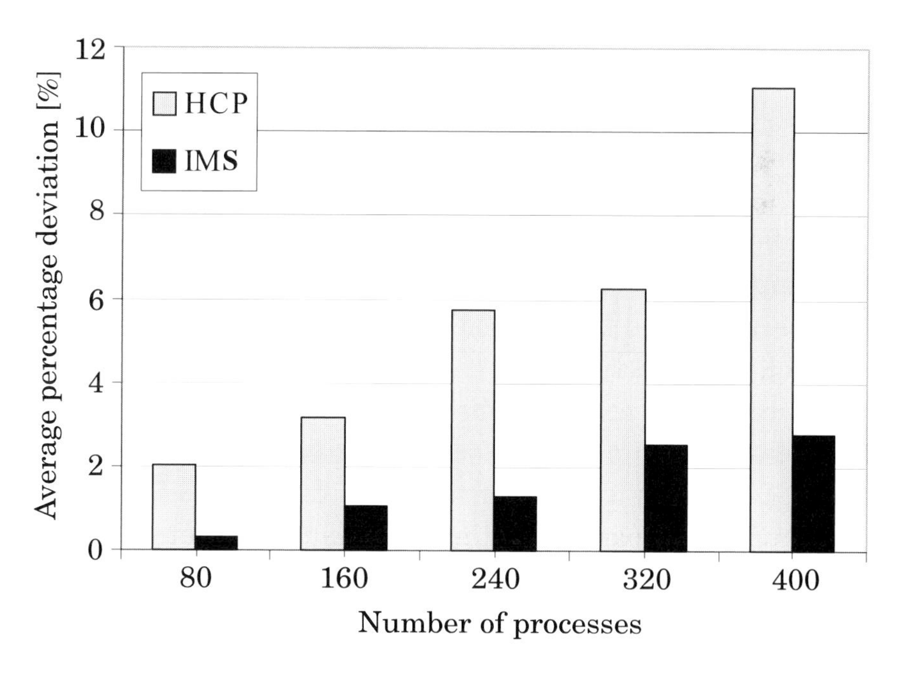

Figure 5.9: Comparison of the IMS and HCP Mapping Heuristics

tion of the total modification cost presented in Section 5.4.3. As a general strategy, we have evaluated our algorithms performing experiments on a large number of test cases generated for experimental purpose. Finally, we have validated the proposed approaches using a real-life example. All experiments were run on a SUN Ultra 10 workstation.

5.5.1 IMS AND THE ITERATIVE DESIGN TRANSFORMATIONS

For the evaluation of our approach we used applications of 80, 160, 240, 320 and 400 processes, representing the application $\Gamma_{current}$, generated for experimental purpose. Thirty applications were generated for each dimension, thus a total of 150 applications were used for experimental evaluation. We considered an architecture consisting of ten nodes of different speeds. For the communication channel we considered a transmission speed of 256 Kbps and a length below 20 meters. The maximum length of the data field in a bus slot was 8 bytes. Throughout the experiments presented in this section we have considered an existing set of applications ψ consisting of 400 processes, with a schedule table of 6 s on each processor, and a slack of about 50% of the total schedule size. In this section we have also considered that no modifications of the existing set of applications ψ are allowed when implementing a new application. We will concentrate on the aspects related to the modification of existing applications, in the following section.

The first result concerns the quality of the designs produced by our initial mapping and scheduling algorithm IMS. As discussed in Section 5.4.1, IMS uses the MPCP priority function which considers particularities of the TDMA protocol. In our experiments we compared the quality of designs (in terms of schedule length) produced by IMS with those generated with the original HCP algorithm proposed in [Jor97]. We have calculated the average percentage deviations of the schedule length produced with HCP and IMS from the length of the best schedule among the two. Results are depicted in Figure 5.9. In average, the deviation from the best result is 3.28 times smaller with IMS than with HCP. The average execution times for both algorithms are under half a second for graphs with 400 processes.

For the next set of experiments we were interested to investigate the quality of the design transformation heuristics discussed in Section 5.4.2, aiming at the optimization of the objective function C. In order to compare this heuristic, implemented in our mapping and scheduling approach MS, we have developed two additional heuristics:

1. A *Simulated Annealing strategy* (SA) (see Appendix A), based on the same moves as described in Section 5.4.2. SA is applied to the solution produced by IMS and aims at finding the near-optimal mapping and schedule that minimizes the objective function C. The main drawback of the SA strategy is that in order to find the near-optimal solution it needs very large computation times. Such a strategy, although useful for the final stages of the system synthesis, cannot be used inside a design space exploration cycle.
2. A so called *Ad-hoc Mapping approach* (AM) which is a simple, straightforward solution to produce designs that, to a certain degree, support an incremental design process. Starting from the initial valid schedule of length S obtained by IMS for a graph G with N processes, AM uses a simple scheme to redistribute the processes inside the $[0, D]$ interval, where D is the deadline of process graph G. AM starts by considering the first process in topological order, let it be P_1. It introduces after P_1 a slack of size max(*smallest process size of* Γ_{future}, $(D - S) / N$), thus shifting all descendants of P_1 to the right (towards the end of the schedule table). The insertion of slacks is repeated for the next process, with the current, larger value of S, as long as the resulted schedule has a length $S \leq D$. Processes are moved only as long as their individual deadlines (if any) are not violated.

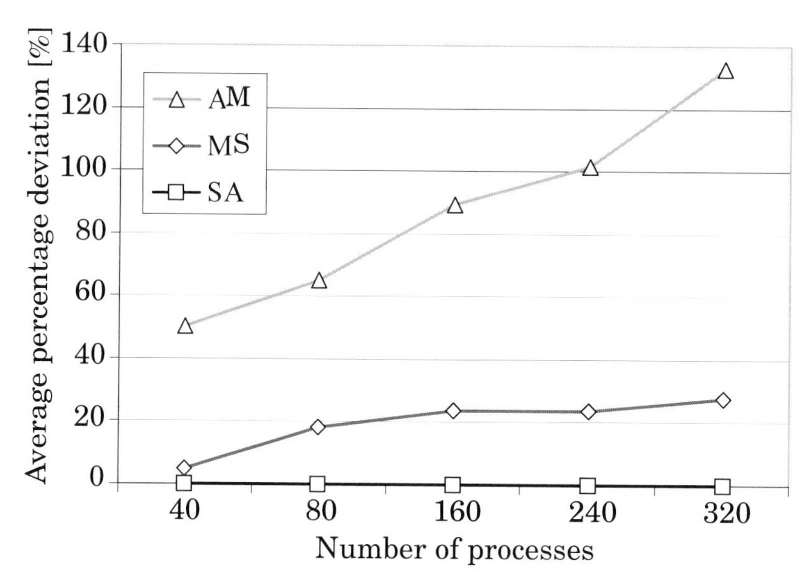

a) Deviation of the objective function obtained with MS and AM from that obtained with SA

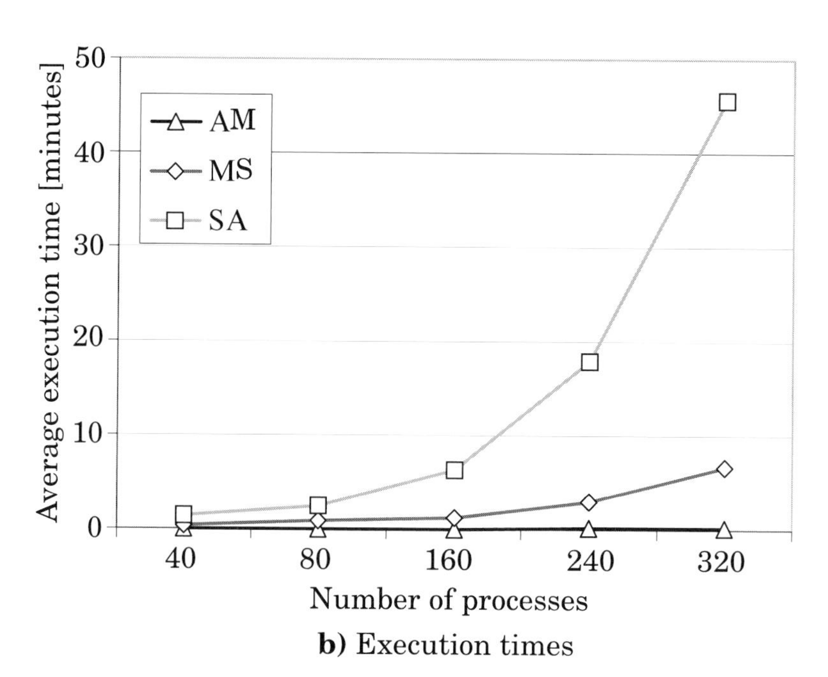

b) Execution times

Figure 5.10: Evaluation of the Design Transformation Heuristics

Our heuristic (MS), proposed in Section 5.4.2, as well as SA and AM have been used to map and schedule each of the 150 process graphs on the target system. For each of the resulted designs, the objective function C has been computed. Very long and expensive runs have been performed with the SA algorithm for each graph and the best ever solution produced has been considered as the near-optimum for that graph. We have compared the objective function obtained for the 150 process graphs considering each of the three heuristics. Figure 5.10a presents the average percentage deviation of the objective function obtained with the MS and AM from the value of the objective function obtained with the near-optimal scheme (SA). We have excluded from the results in Figure 5.10a, 37 solutions obtained with AM for which the second design criterion has not been met, and thus the objective function has been strongly penalized. The average runtimes of the algorithms are presented in Figure 5.10b. The SA approach performs best in terms of quality at the expense of a large execution time: The execution time can be up to 45 minutes for large graphs of 400 processes. The important aspect is that MS performs very well, and is able to obtain good quality solutions, very close to those produced with SA, in a very short time. AM is, of course, very fast, but since it does not address explicitly the two design criteria presented in Section 5.3 it has the worst quality of solutions, as expressed by the objective function.

The most important aspect of the experiments is determining to which extent the design transformations proposed by us, and the related heuristic, really facilitate the implementation of future applications. To find this out, we have mapped applications of 80, 160, 240 and 320 nodes representing the $\Gamma_{current}$ application on top of ψ (the same ψ as defined for the previous set of experiments). After mapping and scheduling each of these graphs we have tried to add a new application Γ_{future} to the resulted system. Γ_{future} consists of a process graph of 80 processes, randomly generated according to the following specifications: S_t={20, 50, 100, 150, 200 ms}, $f_{S_t}(S_t)$ = {0.1, 0.25, 0.45, 0.15, 0.05}, S_b = {2, 4, 6, 8 bytes}, $f_{S_b}(S_b)$ = {0.2, 0.5, 0.2, 0.1}, T_{min} = 250 ms, t_{need} = 100 and b_{need} = 20 ms.

The experiments have been performed two times, using first MS and then AM for mapping $\Gamma_{current}$. In both cases we were interested if it is possible to find a correct implementation for Γ_{future} on top of $\Gamma_{current}$ using the initial mapping and scheduling algorithm IMS (without any modification of ψ or $\Gamma_{current}$). Figure 5.11 shows the percentage of successful implementations of Γ_{future} in the two cases. In the case $\Gamma_{current}$ has been implemented with MS, this means using the design criteria and metrics proposed in this chapter, we were able to find a valid schedule for 65% of the total cases. However, using AM to map $\Gamma_{current}$, has led to a situation where IMS is able to find correct solutions in only 21% of the cases. Another conclusion from Figure 5.11 is that when the total slack available is large, as in the case $\Gamma_{current}$ has only 80 processes, it is easy for MS and, to a certain extent, even for AM to find a mapping that allows adding future applications. However,

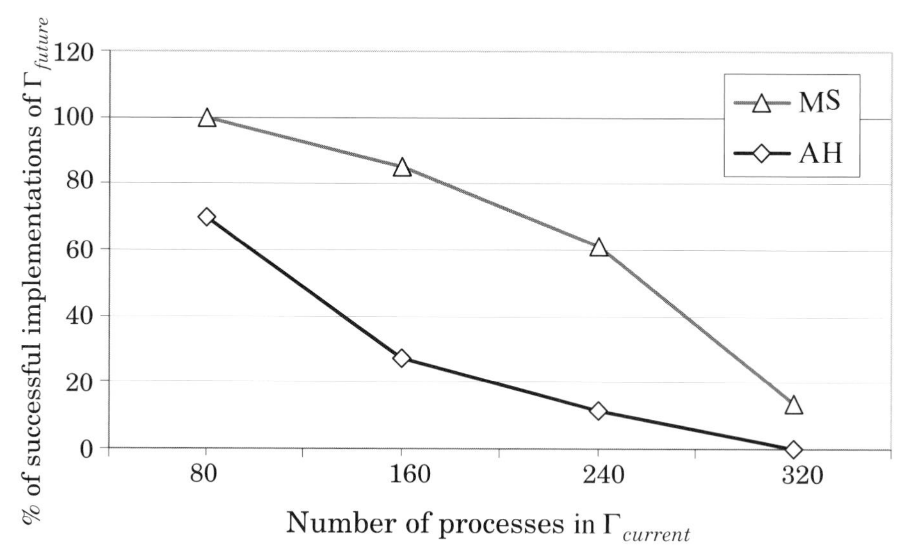

Figure 5.11: Percentage of Future Applications Successfully Implemented

as $\Gamma_{current}$ grows to 240 processes, only MS is able to find an implementation of $\Gamma_{current}$ that supports an incremental design process, accommodating the future application in more then 60% of the cases. If the remaining slack is very small, after we map a $\Gamma_{current}$ of 320 processes, it becomes practically impossible to map new applications without modifying the current system.

5.5.2 MODIFICATION COST MINIMIZATION HEURISTICS

For this set of experiments we first used the same 150 applications as in the previous section, consisting of 80, 160, 240, 320 and 400 processes, for the application $\Gamma_{current}$. We also considered the same system architecture as presented there.

The first results concern the quality of the solution obtained with our mapping strategy MS using the search heuristic SH compared to the case when the simple greedy approach AS and the exhaustive search ES are used. For the existing applications we have generated five different sets ψ consisting of different numbers of applications and processes, as follows: 6 applications (320 processes), 8 applications (400 processes), 10 applications (480 processes), 12 applications (560 processes), 14 applications (640 processes). Each application had an associated modification cost, assigned manually, in the range 10 to 100. The available slack is of about 50% of the total schedule size. The dependencies between applications (in the sense introduced in Section 2.3.2) were such that the total number of possible subsets Ω resulted for each set ψ were 32, 128, 256, 1024, and 4096, respectively. We have considered that the future applications, Γ_{future}, are characterized by the following parameters: S_t = {20, 50, 100, 150, 200 ms}, $f_{S_t}(S_t)$ = {0.1, 0.25, 0.45, 0.15, 0.05}, S_b = {2, 4, 6, 8 bytes}, $f_{S_b}(S_b)$ = {0.2, 0.5, 0.2, 0.1}, T_{min} = 250 ms, t_{need} = 100 ms and b_{need} = 20 ms.

MS has been used to produce a valid solution for each of the 150 applications representing $\Gamma_{current}$, on each of the target configurations ψ using the ES, AS and SH approaches to subset selec-

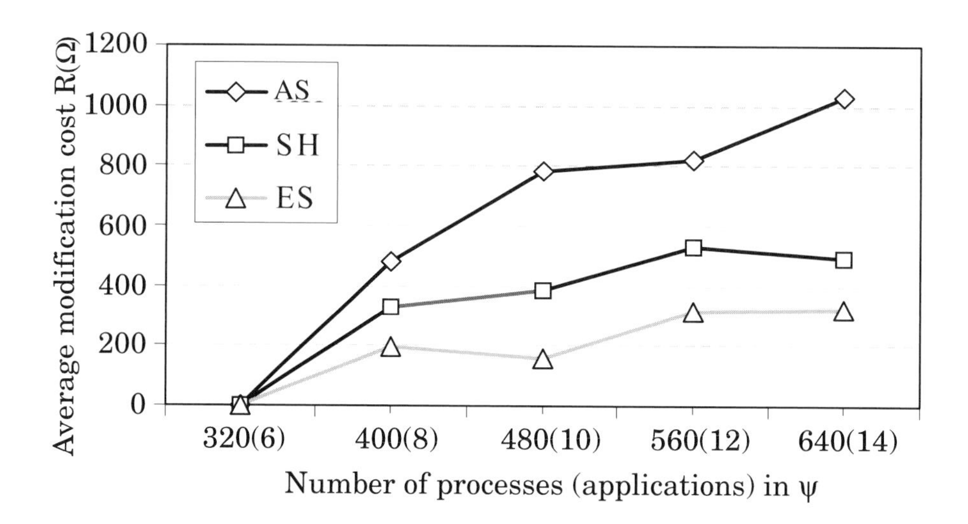

a) Modification cost obtained with the AS, SH, and ES heuristics

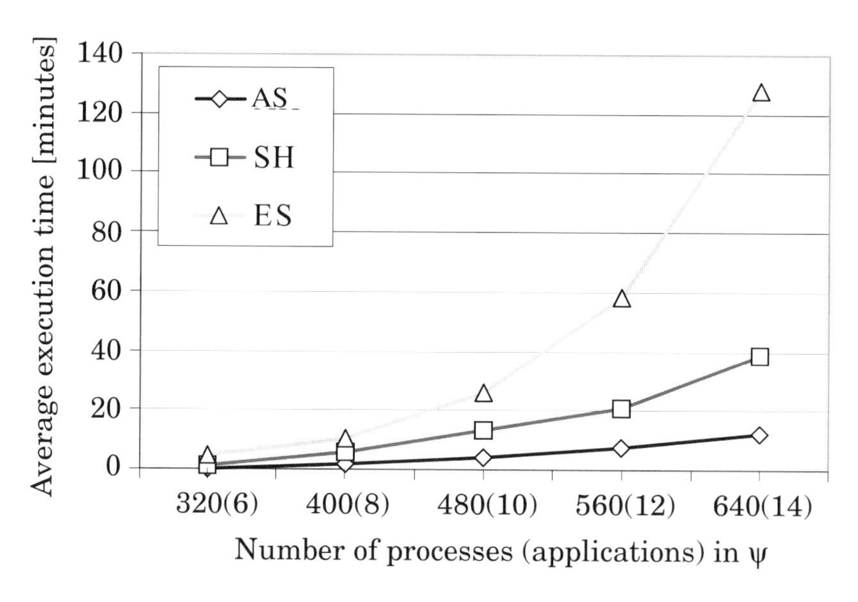

b) Execution times

Figure 5.12: Evaluation of the Modification Cost Minimization Heuristics

tion. Figure 5.12a compares the three approaches based on the total modification cost needed in order to obtain a valid solution. The exhaustive approach ES is able to obtain valid solutions with an optimal (smallest) modification cost, while the greedy approach AS produces in average 3.12 times more costly modifications in order to obtain valid solutions. However, in order to find the optimal solution, ES needs large computation times, as shown in Figure 5.12b. For example, it can take more than 2 hours in average to find the smallest cost subset to be remapped that leads to a valid solution in the case of 14 applications (640 processes). We can see that the proposed heuristic SH performs well, producing close to optimal results with a good scaling for large application sets. For the results in Figure 5.12 we have eliminated those situations in which no valid solution could be produced by MS.

Finally, we have repeated the last set of experiments discussed in the previous section (the experiments leading to the results in Figure 5.11). However, in this case, we have allowed the current system (consisting of $\psi \cup \Gamma_{current}$) to be modified when implementing Γ_{future}. If the mapping and scheduling heuristic is allowed to modify the existing system then we are able to increase the total number of successful attempts to implement application Γ_{future} from 65% to 77.5%. For the case with $\Gamma_{current}$ consisting of 160 processes (when the amount of available resources for Γ_{future} is small) the increase is from 60% to 92%. Such an increase is, of course, expected. The important aspect, however, is that it is obtained not by randomly selecting old applications to be modified, but by performing this selection such that the total modification cost is minimized.

5.5.3 THE VEHICLE CRUISE CONTROLLER

As a real-life case study, we have considered the cruise controller (CC) presented in Section 2.3.3 in order to evaluate our approaches. For the cruise controller we have used:

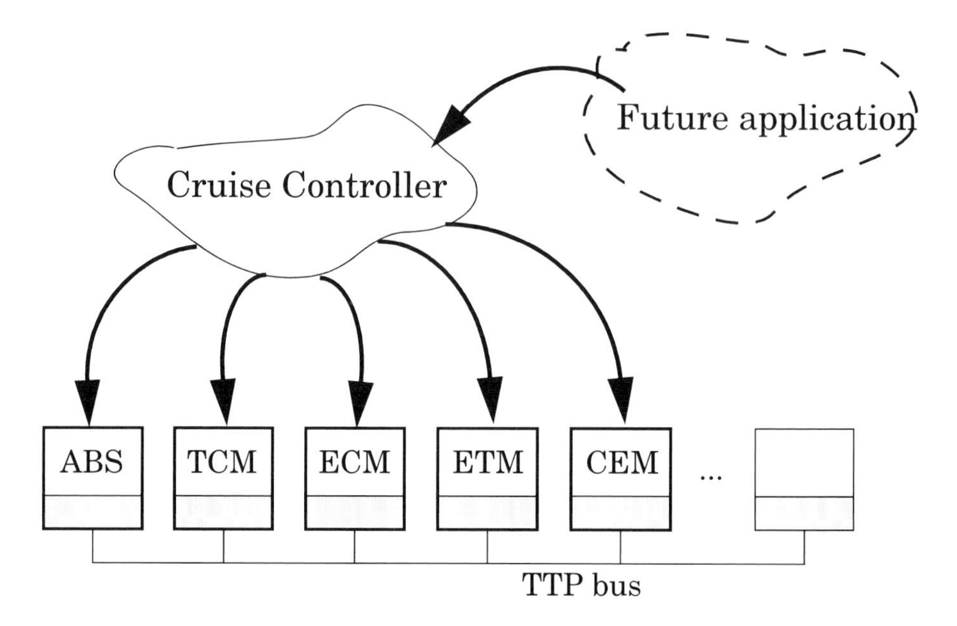

Figure 5.13: Implementation of the Cruise Controller

- the un-mapped model of the cruise controller presented in Figure 2.9 on page 40, with 32 processes and two conditions,
- the hardware architecture in Figure 2.7a on page 37, consisting of five nodes interconnected using a bus implementing the time-triggered protocol,
- and the software architecture for time-driven systems introduced in Section 3.3.
- We have considered a transmission speed of the communication channel of 256 Kbps and the frequency of the TTP controller was chosen to be 20 MHz.
- The period of the CC was chosen to be 300 ms, equal to the deadline.

The system ψ representing the applications already running on the four nodes, has been modeled as a set of 80 processes with a schedule table of 300 ms and leaving a total of 40% slack. The CC is the $\Gamma_{current}$ application to be implemented. We have also generated 30 future applications of 40 processes each, with the general characteristics close to those of the CC, which are typical for automotive applications. We have first mapped and scheduled the CC on top of ψ using the ad-hoc strategy (AM) and then our MS algorithm. On the resulted systems, consisting of $\psi \cup$ CC, we tried to implement each of the 30 future applications. First, we considered a situation in which no modifications of the existing system are allowed when implementing the future applications. In this case, we were able to implement 21 of the 30 future applications after implementing the CC with MS, while using AM to implement the CC, only 4 of the future applications could be mapped. When modifications of the current system were allowed, using MS, we were able to map 24 of the 30 future applications on top of the CC.

As our experiments have shown, the design criteria proposed in this chapter are able to guide our mapping and scheduling approaches to implementations which support an incremental design process. This means that the modifications performed to the existing applications are minimized, and that new functionality, later to be added, can be easily accommodated.

The next part of the book will address the mapping and scheduling in the context of event-driven systems.

PART III
Event-Driven Systems

Chapter 6
Schedulability Analysis and Bus Access Optimization for Event-Driven Systems

IN THE PREVIOUS part of the book we have addressed the issue of non-preemptive static process scheduling and communication synthesis using the TTP as the communication infrastructure.

In the third part of the book, consisting of this and the next chapter, we consider event-driven distributed real-time systems where the activation of processes is event-triggered, while the communications are time-triggered, according to the TTP.

This chapter is structured as follows. The next section presents background and related work in the area of schedulability analysis. In Section 6.2 we go into some details concerning the particular schedulability analysis technique that is used as a starting point in our later discussions. Section 6.3 presents the schedulability analysis we have developed for systems with both control and data dependencies modeled as a set of conditional process graphs. Section 6.4 shows how the current state-of-the-art schedulability analysis for distributed real-time systems can be extended to consider the time triggered protocol. Once realistic communication aspects are captured by the schedulability analysis, they can be used to drive the communication synthesis process described in Section 6.7. Finally, Section 6.8 presents the experimental results obtained for the approaches presented in this chapter.

6.1 Background

Preemptive scheduling of independent processes with static priorities running on single-processor architectures has its roots in the work of Liu and Layland [Liu73]. The approach has been later extended to accommodate more general computational models and has also been applied to distributed systems [Tin94a]. The reader is referred to [Aud95], [Bal98], [Sta93] for surveys on this topic.

In [Yen97] performance estimation is based on a preemptive scheduling strategy with static priorities using rate monotonic analysis. In [Lee99] an earlier deadline first strategy is used for non-preemptive scheduling of processes with possible data dependencies. Preemptive and non-preemptive static scheduling are combined in the co-synthesis environment described in [Dav98], [Dav99].

In many of the previous scheduling approaches researchers have assumed that processes are scheduled independently. However, this is not the case in reality, where process sets can exhibit both data and control dependencies. Moreover, knowledge about these dependencies can be used in order to improve the accuracy of schedulability analyses and the quality of the produced schedules.

One way of dealing with data dependencies between processes with static priority based scheduling has been indirectly addressed by the extensions proposed for the schedulability analysis of distributed systems through the use of the *release jitter* [Tin94a]. Release jitter is the worst case delay between the arrival of a process and its release (when it is placed in the ready-queue for the processor) and can include the communication delay due to the transmission of a message on the communication channel.

In [Tin94b] and [Yen98] time *offset* relationships and *phases*, respectively, are used in order to model data dependencies. Offset and phase are similar concepts that express the existence of a fixed interval in time between the arrivals of sets of processes. The authors show that by introducing such concepts into the computational model, the pessimism of the analysis is significantly reduced when bounding the time behavior of the system. The concept of *dynamic offsets* has been later introduced in [Pal98] and used to model data dependencies [Pal99].

When control dependencies exist then, depending on conditions, only a subset of the set of processes is executed during an invocation of the system. *Modes* have been used to model a certain class of control dependencies [Foh93]. Such a model basically assumes that at the starting of an execution cycle, a particular functionality is known in advance and is fixed for one or several cycles until another mode change is performed. However, modes cannot handle fine grained control dependencies, or certain combinations of data and control dependencies. Careful modeling using the *periods* of processes (lower bound between subsequent re-arrivals of a process) can also be a solution for some cases of control dependencies [Ger96]. If, for example, we know that a certain set of processes will only execute every second cycle of the system, we can set their periods to the double of the period of the rest of the processes in the system. However, using the worst case assumption on periods leads very often to unnecessarily pessimistic schedulability evaluations. More refined process models can produce much better schedulability results, as will be later shown in the book. Recent works [Bar98a], [Bar98b] aim at extending the existing models to handle control dependencies. In [Bar98b] Baruah introduces the *recurring real-time task model* that is able to capture lower level control dependencies, and presents an exponential-time analysis for uniprocessor systems.

As mentioned in Section 4.1, researchers have initially ignored communication aspects when analyzing real-time systems. However, we have to mention here some results obtained in extending real-time schedulability analysis so that network communication aspects can be handled. In [Tin95], for example, the CAN protocol is investigated while the work reported in [Erm97] considers systems based on the ATM protocol. Analysis for a simple time-division multiple access (TDMA) protocol is provided in [Tin94a] that integrates processor and communication schedulability and provides a "holistic" schedulability analysis in the context of distributed real-time systems. Other protocols have also been considered, like the Token Ring [Str89], and the FDDI network architecture [Agr94].

The problem of how to allocate priorities to a set of distributed processes is discussed in [Gut95]. Their priority assignment heuristic is based on the schedulability analysis from [Tin94a].

In this third part of the book we consider the time-triggered protocol (TTP) [Kop03] as the communication infrastructure for a distributed real-time system. However, the research presented is also valid for any other TDMA-based bus protocol that schedules the messages statically based on a schedule table like, for example, the SAFEbus [Hoy92] protocol used in the avionics industry.

6.2 Response Time Analysis

In this part of the book, we consider that processes are scheduled according to a *fixed-priority preemptive scheduling* policy (FPS). This is the most widely used preemptive scheduling approach, whereby each process has a fixed (static) priority which is computed off-line. The processes ready for execution are then executed according to their priority.

6.2.1 BASIC CONCEPTS

The aim of a schedulability analysis is to determine *sufficient* and *necessary* conditions under which an application is schedulable. An application is *schedulable* if there exists at least one scheduling algorithm that is able to produce a feasible schedule. A schedule is *feasible* if all processes can be completed within the specified constraints.

There are basically two approaches to the schedulability analysis in the context of fixed-priority preemptive scheduling: utilization-based tests, and response-time analysis. The *utilization tests* for FPS [Liu73], [Bin01], [Leh89] are not exact (i.e., are only necessary, but not both necessary and sufficient), and/or are not applicable to a more general process model, as we will introduce below.

Thus, in this book we will use a *response time analysis* [Aud91] approach in order to check the exact feasibility of a set of processes. The approach has two steps:

1. In the first step, the analysis derives the worst-case response time of each process (the time it takes from the moment is ready for execution, until it has finished executing).
2. The second step compares the worst case response time of each process to its deadline and, if the response times are smaller or equal to the deadlines, the system is schedulable.

Before going into the details of the response time analysis, let us present the basic concepts we will use, illustrated also in Figure 6.1 (in addition to those introduced in Section 2.3.1 for the application model):

- *Arrival time*, a_i: the time when a process P_i becomes ready for execution. Also known as *request time* or *release time*.
- *Start time*: the time when a process starts its execution.
- *Finishing time*: the time when a process finishes its execution.
- *Response time*, r_i: the time it takes from the arrival of the process P_i, until it finishes executing.
- *Interference*, I_i: the time a process P_i is interrupted by higher priority processes during its execution.
- *Blocking time*, B_i: the time a process P_i has to wait for lower priority processes that are in their critical section and cannot be interrupted.
- *Release jitter*, J_i: the delay between the arrival of process P_i and the start of its execution.
- *Offset*, O_i: the earliest possible arrival time of process P_i, relative to the start of the schedule (also known as *phase*).
- *Transmission delay*, C_m: the time it takes for a message m to reach the destination controller, once it has been sent on the bus (also known as *propagation delay*).
- *Queuing delay*, W_m: is the delay experienced by m at the communication controller, from the time it was produced by the sender process, until is being sent.
- *Communication delay*, r_m: is the time it takes for a message m to reach the desalination process, from the moment it has been produced by the sender process. It is also known as the *end-to-end* communication delay, or *response time* (similar, conceptually, to the response time of a process).

6.2.2 RESPONSE TIME ANALYSIS OVERVIEW

This section presents an overview of the response time analysis used in this book. We start with the basic response time analysis, as outlined in [Aud91]. In the next sections, we extend this analysis for applications with data and control dependencies, implemented on distributed architectures.

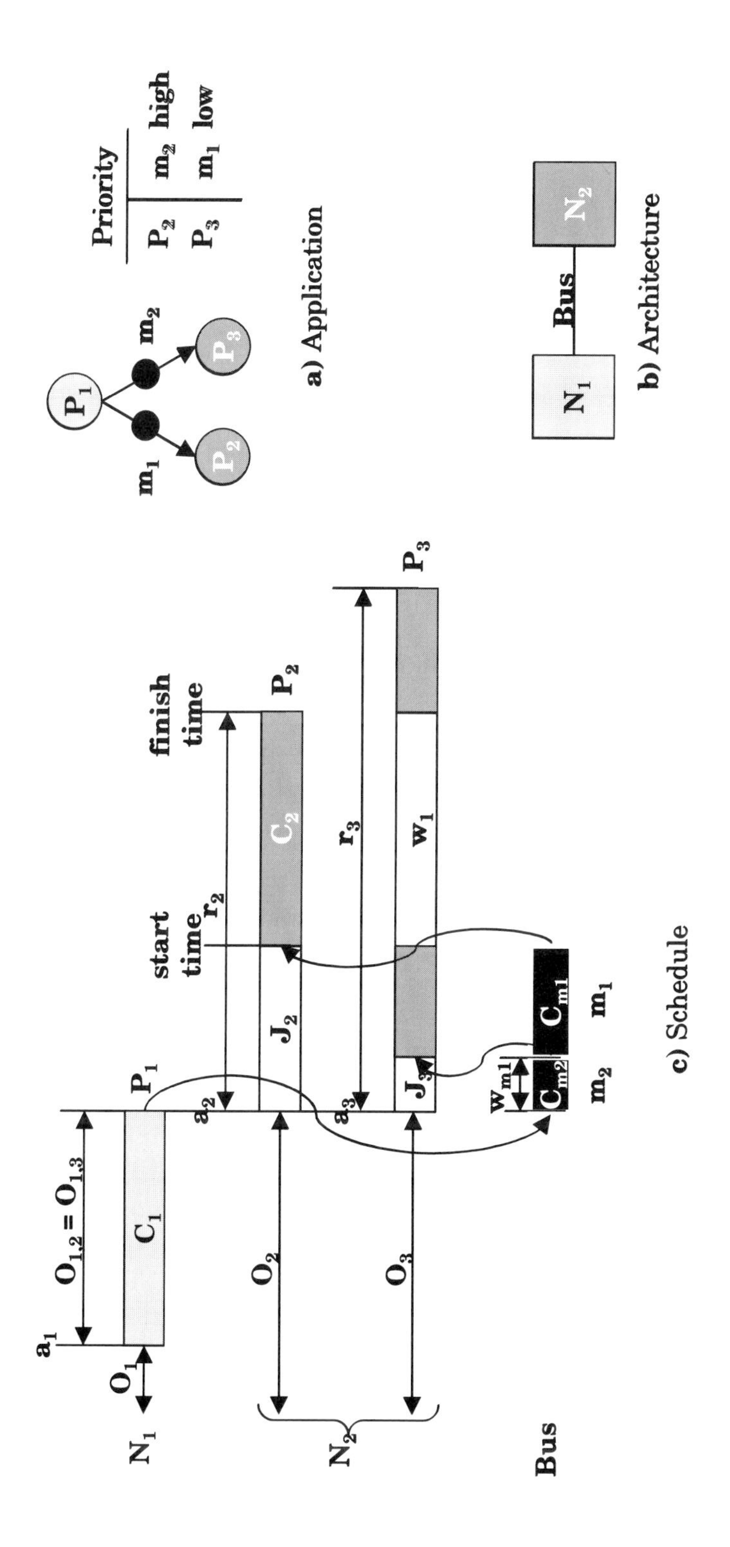

Figure 6.1: Illustration of Schedulability Concepts

Figure 6.2 presents an overview of the schedulability analysis techniques proposed in this chapter (the analyses in the grey boxes are our contribution). Basically, there are two approaches to extending the schedulability analysis. The first category, presented in sections 6.2.4 and 6.3, have focused on reducing the pessimism of the analysis by using the information related to the data and control dependencies, respectively. Sections 6.4 and 6.5 constitute the second category, which has extended the analysis to handle distributed architectures, and the particularities of TTP and CAN protocols.

The analyses presented are structured as follows:

- Section 6.2.3 presents, as mentioned, the basic response time analysis for calculating the worst-case response time r_i of a process P_i. This analysis does not take into account the data and control dependencies that can exist between processes, and it is applicable only to uni-processor systems.
- Section 6.2.4 extends the previous analysis using the information about data dependencies, captured by the offsets, in order to reduce the pessimism[1] of the analysis. Together with the analysis, in Section 6.2.4 we also present an algorithm (DelayEstimate in Figure 6.3) that derives values for offsets such that the schedulability of the application is improved.
- Section 6.3 considers conditional process graphs, that capture not only the dataflow but also the flow of control. The analysis in the Section 6.2 is extended to take into account the information related to the conditions, captured by a CPG, with the aim of reducing the pessimism of the analysis.
- Section 6.4 further extends the analysis to consider applications mapped on distributed architectures. It does this by considering that the release jitter J_i of a receiver process P_i depends on the communication delay of the incoming mes-

1. An analysis A is less pessimistic than an analysis B if it indicates that an application, considered by B not to be schedulable, is, in fact, schedulable.

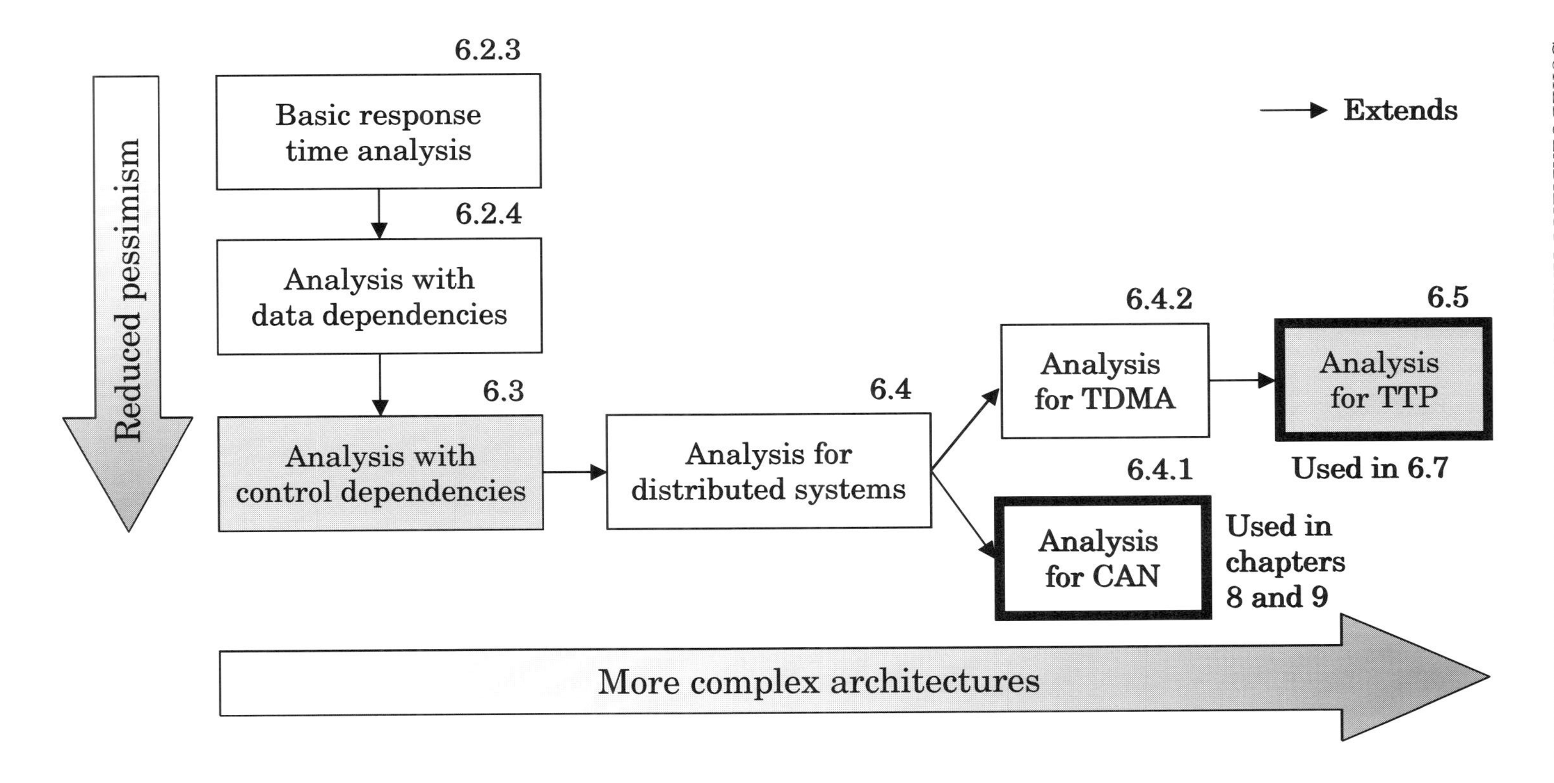

Figure 6.2: Overview of the Schedulability Analysis Approaches

sage m (also called response time r_m of message m). In addition, in Section 6.4 we show how the details of a communication protocol have to be considered when determining the communication delay of a message. In particular, we present the extensions for a simple TDMA protocol and for the CAN bus. For each protocol, we calculate differently the transmission delay C_m, and the worst-case queuing delay W_m of a message m, needed to determine the communication delay, as expressed by Equation 6.6 on page 161.

- Section 6.5 presents the analysis we have developed for the time-triggered protocol. It builds on, and extends, the analysis for a simple TDMA protocol presented in Section 6.4. Four approaches to the scheduling of event-triggered messages over the static TTP bus are presented, with their corresponding analysis for the communication delay (implying deriving, for each case, C_m and W_m).

6.2.3 Basic Response Time Analysis

As mentioned earlier, in order to find out if an application is schedulable, a response time analysis determines the worst-case response time of each process, and then compares it to its deadline. If all response times are smaller than or equal to the deadlines, then the application is schedulable.

Thus, the response time analysis in [Aud91] uses the following equation for determining the worst-case response time r_i of a process P_i:

$$r_i = C_i + \sum_{\forall P_j \in hp(P_i)} \left\lceil \frac{r_i}{T_j} \right\rceil C_j \qquad (6.1)$$

where C_i is the worst-case execution time of process P_i, T_j is the period of process P_j, and $hp(P_i)$ denotes the set of processes that have a priority higher than the priority of P_i.

The summation term, representing the interference I_i of higher priority processes on P_i, increases monotonically in r_i, thus solutions can be found using a recurrence relation. Moreover, the recurrence relations that calculate the worst case response time are guaranteed to converge if the processor utilization is under 100%.

The previously presented analysis assumes that the deadline of a process is smaller or equal to its period. This assumption has later been relaxed [Tin94a] to consider *arbitrary deadlines* (i.e., deadlines can be larger than the period) and *release jitter* (the delay between the arrival of a process and the start of its execution). Thus, the worst-case response time r_i of a process P_i becomes:

$$r_i = \max_{q = 0, 1, 2\ldots} (J_i + w_i(q) - qT_i) \quad (6.2)$$

where q is the number of busy periods being examined, and $w_i(q)$ is the width of the level-i busy period starting at time qT_i. The level-i busy period is defined as the maximum time a processor executes processes of priority greater than or equal to the priority of process P_i, and is calculated as [Tin94a]:

$$w_i(q) = (q+1)C_i + B_i + \sum_{\forall P_j \in hp(P_i)} \left\lceil \frac{w_i(q) + J_j}{T_j} \right\rceil C_j . \quad (6.3)$$

The schedulability analyses presented in the rest of the book work under the following assumptions:

- All the processes belonging a process graph G have the same period T_G. For processes with different periods, a hypergraph is constructed having the LCM of the periods of the processes involved. However, process graphs can have different periods.
- The offsets (see next section) are *static* (as opposed to *dynamic* [Pal98]), and are smaller than the period.
- The deadlines are arbitrary and can be larger than the periods.

6.2.4 SCHEDULABILITY ANALYSIS WITH DATA DEPENDENCIES

The pessimism of the previous analysis can be reduced by using the information related to the precedence relations between processes. The basic idea is to exclude certain worst case scenarios, from the critical instant analysis, which are impossible due to precedence constraints.

Methods for schedulability analysis of data dependent processes with static priority preemptive scheduling have been proposed in [Yen98], [Tin94b], [Pal98], [Pal99]. They use the concept of *offset* (or *phase*), in order to handle data dependencies. [Tin94b] shows that the pessimism of the analysis is reduced through the introduction of offsets. The offsets have to be determined by the designer.

In their analysis [Tin94b], the response time of a process P_i is:

$$r_i = \max_{q = 0, 1, 2...}\left(\max_{\forall P_j \in G}\left(w_i(q) + O_j + J_j - T_G\left(q + \left\lceil \frac{O_j + J_j - O_i - J_i}{T_G} \right\rceil \right) - O_i \right)\right) \quad (6.4)$$

where T_G the period of the process graph G, O_i and O_j are offsets of processes P_i and P_j, respectively, and J_i and J_j are the release jitters of P_i and P_j. In Equation (6.4), the level-i busy period starting at time qT_G [Tin94b] is:

$$w_i(q) = (q + 1)C_i + B_i + I_i. \quad (6.5)$$

In the previous equation, the blocking term B_i represents interference from lower priority processes that are in their critical section and cannot be interrupted, and C_i represents the worst-case execution time of process P_i. The last term captures the interference I_i from higher priority processes in the application, including higher priority processes from other process graphs. The reader is directed to [Tin94b] for the details of the interference calculation.

Although this analysis is exact (both necessary and sufficient), it is computationally infeasible to evaluate. Hence, [Tin94b] proposes a feasible[1] but not exact analysis (sufficient but not necessary) for solving Equation (6.4). Our implementations use the feasible analysis provided in [Tin94b] for deriving the worst-case response time of a process P_i.

The authors in [Yen98] provide a framework that iteratively finds the phases (offsets) for all processes, and then feeds them back into the schedulability analysis which in turn is used again to derive better phases. Thus, the pessimism of the analysis is iteratively reduced.

Response Time Analysis Algorithm

In [Yen98] an application is modeled as a set S of n process graphs τ_i, $i = 1, 2, ..., n$. The application model assumed and the definition of a process graph is similar to our CPG, but without considering any conditions. The aim of the schedulability analysis in [Yen98] is to derive an as tight as possible worst case delay on the execution time of each of the process graphs in the application. This delay estimation is done using the algorithm DelayEstimate described in Figure 6.3.

At the core of this algorithm is a worst case response time calculation based on offsets, similar to the analysis in [Tin94b]. Thus, in the LatestTimes function (called in line 8 of DelayEstimate), worst-case response times and upper bounds for the offsets are calculated, while the EarliestTimes function (line 9) calculates the lower bounds of the offsets.

The LatestTimes function is a modified critical-path algorithm that calculates for each node of the graph the longest path to the sink node (see Section 4.2.2 for the definition of the critical

1. The implementation of this feasible analysis is available at: ftp://ftp.cs.york.ac.uk/pub/realtime/programs/src/offsets/

path). Hence, during the topological traversal of the graph τ within LatestTimes, for each process P_i, the worst case response time r_i is calculated according to the Equation 6.4. This value is based on the values of the offsets known so far. Once an r_i is calculated, it can be used to determine and update offsets for other successor processes. Accordingly, the EarliestTimes function determines the lower bounds on the offsets. The influence on graph τ from other graphs in the application is considered in both of the functions mentioned earlier.

These calculations can be improved by realizing that for a process P_i, there might exist a process P_j mapped on the same pro-

```
DelayEstimate(process graph τ, application S)
1  -- derives the worst case delay of a process graph τ considering
2  -- the influence from all other process graphs in the application S
3  for each pair (Pi, Pj) in τ do
4      maxsep[Pi, Pj] = ∞
5  end for
6
7  repeat
8      LatestTimes(τ)
9      EarliestTimes(τ)
10     for each Pi ∈ τ do
11         MaxSeparations(Pi)
12     end for
13 until maxsep is not changed or limit reached
14 return the worst case delay δτ of the graph τ
end DelayEstimate

SchedulabilityTest(application S)
1  -- derives the worst case delay for each process graph in the system
2  -- and verifies if the deadlines are met
3  for each process graph τi ∈ S do
4      DelayEstimate(τi, S)
5  end for
6  if all process graphs meet their deadline then application S is schedulable
end SchedulabilityTest
```

Figure 6.3: Delay Estimation and Schedulability Analysis for Process Graphs

cessor, with $priority_{P_i}$ < $priority_{P_j}$, such that their execution windows never overlap. In this case, the interference of P_j on the execution of P_i can be dropped, resulting in a tighter worst-case response time calculation. This situation is expressed through the so called maxsep table, computed by the MaxSeparations function, whose value maxsep[P_i, P_j] is less than or equal to 0 if the two processes never overlap during their execution (lines 10–12 of DelayEstimate). The term maxsep stands for *maximum separation*, an analysis modified from [Mc92] which builds the maxsep table based on the worst case execution times and offsets determined in EarliestTimes and LatestTimes.

Having a better view on the maximum separation between each pair of processes, tighter worst case execution times and offsets can be derived, which in turn contribute to the update of the maxsep table. This iterative tightening process is repeated until there is no modification to the maxsep table, or a certain imposed limit on the number of iterations is reached (line 13).

Finally, the DelayEstimate function returns the worst-case delay δ_τ estimated for a process graph τ, as the time when the sink node of τ finishes its execution (line 14). Based on the delays produced by DelayEstimate, the function SchedulabilityTest in Figure 6.3 concludes on the schedulability of the application.

6.3 Schedulability Analysis under Control and Data Dependencies

In the previous sections we were interested to extend the basic schedulability analysis to handle data dependencies.

In this section, we are interested further extend the analysis to handle not only data but also control dependencies. This means developing a schedulability analysis for an application modeled as a set of conditional process graphs.

Example 6.1: To show the relevance of our problem, let us consider the example depicted in Figure 6.4, where we have an application modeled as two conditional process graphs G_1 and G_2 with a total of 9 processes (processes P_0, P_8, P_9 and P_{12} are dummy processes and are not counted), and one condition. The processes are mapped on three different processors as indicated by the shading, and the worst case execution time in milliseconds for each process on its respective processor is depicted to the left of each node. G_1 has a period of 200 ms, G_2 has a period of 150 ms. The deadlines are 100 ms on G_1 and 90 ms on G_2.

Table 6.1 presents the worst-case delays of the two graphs. In the column labelled "no conditions" we have the results for the case when the analysis is applied to the set of processes, ignoring control dependencies. This results in a worst case

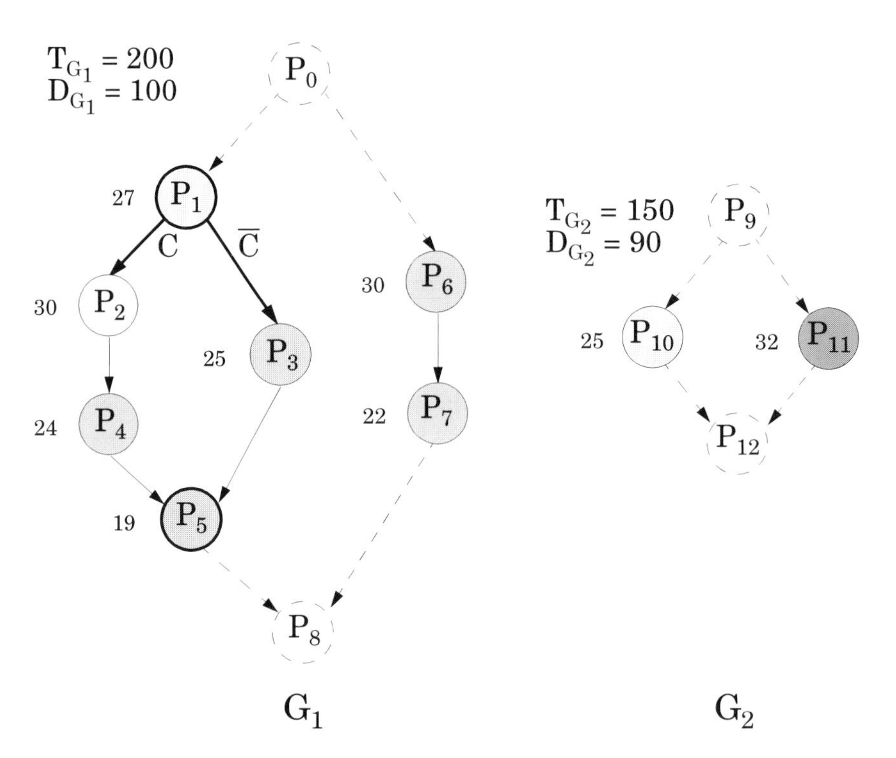

Figure 6.4: Application with Control and Data Dependencies

Table 6.1: Worst-Case Delays for the Application in Figure 6.4

CPG	Worst Case Delays	
	no conditions	conditions
G_1	120	100
G_2	82	82

delay of 120 ms for G_1 and 82 ms for G_2. Hence, the application is considered not to be schedulable. This analysis assumes as a worst case scenario the possible activation of all nine processes, during each execution of the application. This is the solution which will be obtained using a dataflow graph representation of the application.

However, considering the CPG G_1 in Figure 6.4, it is easy to observe that process P_3 on the one side and processes P_2 and P_4 on the other side will not be activated during the same period of G_1. Making use of this information for the analysis we obtain a worst case delay of 100 ms for G_1, as shown in Table 6.1 in the column headed "conditions," which indicates that the application is, in fact, schedulable.

■

Section 2.3.1 has presented the conditional process graph representation. Before introducing our schedulability analysis for CPGs, we reinforce two concepts: the *unconditional subgraphs* and the *process guards*.

Depending on the values calculated for the conditions, different alternative paths through a conditional process graph are activated for a given activation of the application. To model this, a logical expression X_{P_i}, called guard (introduced in Section 2.3.1), can be associated to each node P_i in the graph. It represents the necessary condition for the respective process to be activated.

Example 6.2: In Figure 6.5, for example, $X_{P_4} = C \wedge \overline{D}$, $X_{P_5} = \overline{C}$, $X_{P_9} = true$, $X_{P_{11}} = true$, and $X_{P_{12}} = K$.

■

We call an alternative path through a conditional process graph, resulting from a combination of conditions, an *unconditional subgraph*, denoted by g.

Example 6.3: The CPG G_1 in Figure 6.5 has three unconditional subgraphs, corresponding to the following three combinations of conditions: $C \wedge D$, $C \wedge \overline{D}$, and $\overline{C}$. The unconditional subgraph corresponding to the combination $C \wedge \overline{D}$ in the CPG G_1 consists of processes P_1, P_2, P_4, P_6, P_7, P_9 and P_{10}.

■

The guards of each process, as well as the unconditional subgraphs resulting from a conditional process graph G, can be determined through a simple recursive topological traversal of G.

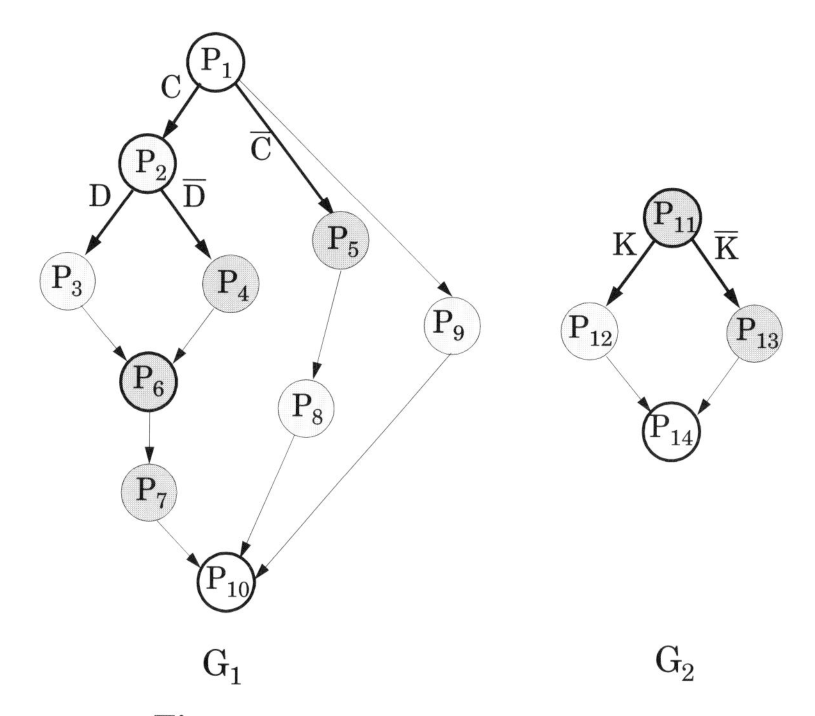

Figure 6.5: Example of Two CPGs

In the following sections we present four approaches to the analysis of conditional process graphs. There are two extreme solutions to this problem:

- The first one, called *Ignoring Conditions* (IC), ignores control dependencies and applies the schedulability analysis for the (unconditional) process graphs.
- At the other end, the *Brute Force Algorithm* (BF) applies the schedulability analysis after each of the CPGs in the application have been decomposed in their constituent unconditional subgraphs.

The other two solutions proposed are in-between solutions:

- *Conditions Separation* (CS) is similar to *Ignoring Conditions*, but uses the knowledge about the conditions in order to update the maxsep table: maxsep[P_i, P_j] = 0 if processes P_i and P_j are on different conditional paths (see Section 6.2.4, Figure 6.3).
- *Relaxed Tightness Analysis* (with two variants: RT1, RT2) is similar to the *Brute Force Algorithm*, but tries to reduce the execution time by removing the iterative tightening loop (hence the name relaxed tightness) in the DelayEstimation function in Figure 6.3.

6.3.1 IGNORING CONDITIONS (IC)

A straightforward approach to the schedulability analysis of applications represented as CPGs is to ignore control dependencies and to apply the schedulability analysis as described in Section 6.2.4 (the algorithm SchedulabilityTest in Figure 6.3).

This means that the conditional edges in the CPGs are considered like simple edges and the conditions in the model are dropped (line 3 of the algorithm in Figure 6.6). What results is an application S consisting of simple process graphs τ_i, each one derived from a CPG G_i of the given application Γ. The application S can then be analyzed (line 4) using the algorithm in

Figure 6.3. It is obvious that if the application S is schedulable, the application Γ is also schedulable (line 5).

This approach, which we call IC, is, of course, very pessimistic. However, this is the current practice when worst-case arrival periods are considered and classical data flow graphs are used for modeling and scheduling [Yen98], [Tin94b].

6.3.2 BRUTE FORCE SOLUTION (BF)

The pessimism of the IC approach can be reduced by considering the conditions captured by a conditional process graph model. A simple, brute force solution is to apply the schedulability analysis presented in Section 6.2.4, after the CPGs have been decomposed into their constituent unconditional subgraphs.

Consider an application Γ which consists of n CPGs G_i, $i = 1, 2, ..., n$. Each CPG G_i can be decomposed into n_i unconditional subgraphs g_j^i , $j = 1, 2, ..., n_i$. In Figure 6.5, for example, we have three unconditional subgraphs g_1^1, g_2^1, g_3^1 derived from G_1 and two, g_1^2, g_2^2 derived from G_2.

At the same time, each CPG G_i can be transformed into a simple process graph τ_i, by transforming conditional edges into ordinary ones and dropping the conditions. When deriving the worst case delay on G_i we apply the analysis from Section 6.2.4 (algorithm DelayEstimate in Figure 6.3) separately to each unconditional subgraph g_j^i in combination with the graphs (τ_1, τ_2, ... τ_{i-1}, τ_{i+1}, τ_n). This means that we consider each alternative path from

```
SA/IC(application Γ)
1  -- verifies the schedulability of a system consisting of a set of
2  -- conditional process graphs
3  transform each G_i ∈ Γ into the corresponding τ_i ∈ S
4  SchedulabilityTest(S)
5  if S is schedulable then application Γ is schedulable
end SA/IC
```

Figure 6.6: Schedulability Analysis Ignoring Conditions

G_i in the context of the application, instead of the whole subgraph τ_i as in the previous approach. This is described by the algorithm DE/CPG in Figure 6.7a. The schedulability analysis is then based on the delay estimation for each CPG as shown in the algorithm SA/BF in Figure 6.7b.

Such an approach, we call it BF, while producing tight bounds on the delays, can be expensive from the runtime point of view, because it is applied for each unconditional subgraph. In general, the number of unconditional subgraphs can grow exponentially. However, for many of the practical systems this is not the

```
DE/CPG(CPG G, application S)
1  --  derives the worst case delay of a CPG G considering
2  --  the influence from all other process graphs in the application S
3  extract all unconditional subgraphs g_j from G
4  for each g_j ∈ G do
5      DelayEstimate(g_j, S)
6  end for
7  return the largest of the delays, which is
       the worst case delay δ_Γ of CPG G
end DE/CPG
```

a) DE/CPG: Delay estimation for conditional process graphs

```
SA/BF(application Γ)
1  --  verifies the schedulability of a system consisting of a set Γ of
2  --  conditional process graphs
3  transform each G_i ∈ Γ into the corresponding τ_i ∈ S
4  for each G_i ∈ Γ do
5      DE/CPG(G_i, {τ_1, τ_2, ...τ_{i-1}, τ_{i+1}, τ_n})
6  end for
7  if all CPGs meet their deadline then the application Γ is schedulable
end SA/BF
```

b) SA/BF: Schedulability analysis: the brute force approach

Figure 6.7: Brute Force Schedulability Analysis

case, and the brute force method can be used. Alternatively, less expensive methods, like those presented next, can be applied.

6.3.3 CONDITION SEPARATION (CS)

In some situations, the explosion of unconditional subgraphs makes the brute force method inapplicable. Hence, we need to find an analysis that is situated somewhere between the two alternatives IC and BF, which means its should not be too pessimistic and should run in acceptable time.

A first idea is to go back to the DelayEstimate algorithm in Figure 6.3, and use the knowledge about conditions in order to update the maxsep table. If two processes P_i and P_j never overlap their execution because they execute under alternative values of conditions, then we can update maxsep[P_i, P_j] to 0, and thus, improve the quality of the delay estimation. Two processes P_i and P_j never overlap their execution if there exists at least one condition C, so that $C \subset X_{P_i}$ (X_{P_i} is the guard of process P_i) and $\overline{C} \subset X_{P_j}$ (lines 19–23 in Figure 6.8).

In this approach, called CS, we practically use the same algorithm as for ordinary process graphs and try to exploit the information captured by conditional dependencies in order to exclude certain influences during the analysis. In Figure 6.8 we show the algorithm SA/CS which performs the schedulability analysis based on this heuristic.

6.3.4 RELAXED TIGHTNESS ANALYSIS (RT)

The two alternatives of the RT approach discussed here are similar to the brute force algorithm in Figure 6.7. However, they try to improve on the execution time of the analyses by reducing the complexity of the DelayEstimate algorithm (Figure 6.3) which is called from the DE/CPG function, in line 5 (Figure 6.7a). This will reduce the execution time of the analysis, not by reducing the

```
SA/CS(application Γ)
1  -- verifies the schedulability of a system consisting of a set Γ of
2  -- conditional process graphs
3  transform each G_i ∈ Γ into the corresponding τ_i ∈ S
4      and keep guard X_{P_i} for each P_i
5  for each τ_i ∈ S do
6      -- derives the worst case delay of a process graph τ_i
7      -- considering the influence from all other process graphs
8      -- in the system S
9      for each pair (P_i, P_j) in τ_i do
10         maxsep[P_i, P_j] = ∞
11     end for
12
13     repeat
14         LatestTimes(τ_i)
15         EarliestTimes(τ_i)
16         for each P_i ∈ τ_i do
17             MaxSeparations(P_i)
18         end for
19         for each pair (P_i, P_j) in τ_i do
20             if ∃ C, C ⊂ X_{P_i} ∧ C̄ ⊂ X_{P_j} then
21                 maxsep[P_i, P_j] = 0
22             end if
23         end for
24     until maxsep is not changed or limit reached
25         δ_{G_i} is the worst case delay for G_i
26 end for
27 if all CPGs meet their deadline then the application Γ is schedulable
end SA/CS
```

Figure 6.8: Schedulability Analysis using Condition Separation

number of subgraphs which have to be visited (like in the CS approach), but by reducing the time needed to analyze each subgraph.

As our experimental results in Section 6.8 show, this approach can be very effective in practice. Of course, by the simplification

```
DelayEstimateRT1(process graph τ, application S)
1  LatestTimes(τ)
end DelayEstimateRT1
```

a) Delay estimation for RT1

```
DelayEstimateRT2(process graph τ, application S)
1  for each pair (Pi, Pj) in τi do
2      maxsep[Pi, Pj] = ∞
3  end for
4  LatestTimes(τ)
5  EarliestTimes(τ)
6  for each Pi ∈ τ do
7      MaxSeparations(Pi)
8  end for
9  LatestTimes(τ)
end DelayEstimateRT2
```

b) Delay estimation for RT2

Figure 6.9: Delay Estimation for the RT Approaches

applied to DelayEstimate the quality of the analysis is reduced in comparison to the brute force method.

We have considered two alternatives of which the first one is more drastic while the second one is trying a more refined trade-off between execution time and quality of the analyses.

With both these approaches, the idea is not to run the iterative tightening loop in DelayEstimate that repeats until no changes are made to maxsep or until the limit is reached (lines 7–13 in Figure 6.3). While this tightening loop iteratively reduces the pessimism when calculating the worst case response times, the actual calculation of the worst case response times is done in LatestTimes, and the rest of the algorithm in Figure 6.3 just tries to improve on these values. For the first approach, called RT1 the function DelayEstimate has been transformed like in Figure 6.9a.

However, it might be worth using at least the MaxSeparations in order to obtain tighter values for the worst case response times. For the alternative RT2 in Figure 6.9b, DelayEstimateRT2 first calls LatestTimes and EarliestTimes, then MaxSeparations in order to build the maxsep table, and again LatestTimes to tighten the worst case response times (lines 4–9 in Figure 6.9b).

6.4 Schedulability Analysis for Distributed Systems

The previous sections have shown how we can reduce the pessimism of the analysis by using information related to the data and control dependencies. In this section, we present an extension of the response time analysis to handle applications distributed on multi-processor architectures.

Tindell et al. [Tin94a] integrate processor and communication scheduling and provide a "holistic" schedulability analysis in the context of distributed real-time systems. The validity of the analysis in has been later confirmed in [Pal97].

In the case of a distributed system the response time of a process also depends on the communication delay due to messages. In [Tin94a] the analysis for messages is done is a similar way as for processes: a message is seen as an un-preemptable process that is "running" on a bus. Thus, the same analysis can be applied for messages on a bus, rewriting Equation 6.1 to:

$$r_m = \max_{q = 0, 1, 2\ldots} (J_m + W_m(q) + C_m), \tag{6.6}$$

where J_m is the jitter of message m which in the worst case is equal to the response time $r_{S(m)}$ of the sender process $P_{S(m)}$ and C_m is the worst-case time it takes for message m to reach the destination controller. W_m is the *worst-case queuing delay* experienced by m at the communication controller, and is calculated as:

$$W_m(q) = w_m(q) - qT_m \tag{6.7}$$

where q is the number of busy periods being examined, and $w_m(q)$ is the width of the level-m busy period starting at time qT_m.

The response time analyses for processes and messages are combined by realizing that the jitter of a destination process depends on the communication delay between sending and receiving a message. Thus, for a process $P_{D(m)}$ that receives a message m from a sender process $P_{S(m)}$, the release jitter is:

$$J_{D(m)} = r_m. \tag{6.8}$$

For the communication infrastructure of our heterogeneous architectures, we use in this book the controller area network and the time-triggered protocols. In order to analyze systems implemented with these two protocols, we have to provide analyses that bound the worst-case queuing delay W_m and worst-case transmission delay C_m for a message m. Tindell et al. [Tin95] provide an analysis for the CAN protocol, while in [Tin94a] an analysis for a simple TDMA protocol, sharing similarities with TTP, is presented. We present briefly these analyses, and later we will show how they can be extended to suit our particular settings.

6.4.1 Schedulability Analysis for the CAN Protocol

Tindell et al. [Tin95] provide worst-case bounds for w_m and C_m in the context of the CAN protocol. CAN is a priority bus, where the message having the highest priority on the network gets to be transmitted (see Section 3.2.2).

In Figure 3.8 on page 60 node N_2 is part of a CAN network, and has a CAN controller. Messages waiting to become the highest priority on the network wait for their transmission in an outgoing queue denoted in the figure with Out_{N_2}. Thus, the worst-case queuing delay W_m for a message m is determined according to Equation (6.7) where the level-m busy period $w_m(q)$ is:

$$w_m(q) = B_m + \sum_{\forall m_j \in hp(m)} \left\lceil \frac{w_m(q) + J_j}{T_j} \right\rceil C_j . \quad (6.9)$$

The intuition is that m has to wait, in the worst case, first for the largest lower priority message that is just being transmitted (B_m) as well as for the higher priority $m_j \in hp(m)$ messages that have to be transmitted ahead of m (the second term). In the worst case, the time it takes for the largest lower priority message $m_k \in lp(m)$ to be transmitted to its destination is:

$$B_m = \max_{\forall m_k \in lp(m)} (C_k) . \quad (6.10)$$

Once m is sent, the time C_m it takes to transmit it to the destination controller depends on the frame configuration introduced in Section 3.2.2, message size s_m, and the time τ_{bit} it takes to transmit a bit [Tin95]:

$$C_m = \left(\left\lfloor \frac{34 + 8s_m}{5} \right\rfloor + 47 + 8s_m \right) \tau_{bit} \quad (6.11)$$

6.4.2 SCHEDULABILITY ANALYSIS FOR A TDMA BUS

In this part of the book we consider that, although the messages are produced based on events, they are transmitted using the time-triggered protocol. TTP has a time-division multiple access scheme to the bus, meaning that a message produced on node N_i can be transmitted only during a predetermined time interval, the slot S_i corresponding to node N_i.

Tindell et al. [Tin94a] have developed an analysis for a simple TDMA bus that share similarities with the TTP. In their setting, messages are split into packets before being sent.

In this case, the transmission of a message can span over several rounds. The analysis in [Tin94a] determines the transmission delay C_m based on the number of packets that need to be taken from the queue over the time $w_m(q)$ in order to guarantee the transmission of the last packet of m. Thus, the propagation delay depends on the number of busy periods q being examined.

Thus, the transmission delay C_m of a message m sent as p_m packets over a slot S, is equal to:

$$C_m = \left\lceil \frac{p_m}{S_p} \right\rceil S_S \tau_{bit}, \tag{6.12}$$

where S_p is the size of slot S in number of packets and S_S is the size of the slot in number of bits.

In the case when messages cannot be split into packets and thus the transmission of a message is done in one single round, the propagation delay C_m is equal to the slot size in which message m is being transmitted, and thus is not influenced by the number of busy periods being examined.

The details of message transmission in such a setting are presented in Section 3.4. When a message is produced by a sender process, all its packets are placed in the *Out* queue (Figure 3.6 on page 56). Packets are ordered according to their priority. At its activation, the message transfer process takes a certain number of packets from the head of the *Out* queue and constructs a frame. The number of packets accepted is decided so that their total size does not exceed the length of the data field of the frame. This length is limited by the size of the slot corresponding to the respective processor. Since the messages are produced dynamically, they have to be identified in a certain way so that they are recognized when the frame arrives at the delivery process. Thus, each message has several identifier bits appended at the beginning of the message.

Since the packets are dynamically packed into frames in the order they are sorted in the queue, the worst-case queuing delay to the communication channel for a packet p depends on the number of packets queued ahead of it.

The analysis in [Tin94a] bounds the number of queued ahead *packets* of messages of higher priority than a message m by:

$$w_m(q) = \left\lceil \frac{(q+1)p_m + I_m(w_m(q))}{S_p} \right\rceil T_{TDMA}, \qquad (6.13)$$

where p_m is the number of packets of message m, S_p is the size of the slot (in number of packets) corresponding to m, and

$$I_m(w_m(q)) = \sum_{\forall m_j \in hp(m)} \left\lceil \frac{w_m(q) + r_{S(m_j)}}{T_j} \right\rceil p_j, \qquad (6.14)$$

where p_j is the number of packets of a message m_j.

The analysis assumes that the period T_m of any message m is longer or equal to the length of a TDMA round, $T_m \geq T_{TDMA}$ (see Figure 3.2 on page 48 and Figure 6.10 on page 167).

6.5 Schedulability Analysis for the Time Triggered Protocol

Although there are many similarities with the general TDMA protocol, the analysis in the case of TTP is different in several aspects from the one outlined in the previous sections, and also differs to a large degree depending on the policy chosen for message scheduling.

The four approaches we propose for scheduling of messages using TTP differ in the way the messages are allocated to the communication channel (either statically or dynamically), and whether they are split into packets for transmission or not. The next sections present the analysis for each approach as well as the degrees of liberty a designer has, in each of the cases, for optimizing the MEDL (the static schedule table for messages, see Section 3.2.1).

Before going into details for each of the message scheduling approaches proposed by us we have to mention how we account on each processors, for the overheads due to transmission.

The overhead produced by the communication activities must be accounted for not only as part of the response time for a message, but also through its influence on the response time of processes running on the same processor. We consider this influence during the schedulability analysis of processes on each processor. We assume that the worst case computation time of the transfer process (T in Figure 3.6 on page 56) is known, and that it is different for each of the four message scheduling approaches. Based on the respective MHTT, the transfer process is activated for each frame sent. Its worst-case period is derived from the minimum time between successive frames.

The response time of the delivery process (D in Figure 3.6), is considered as part of the communication delay. The influence due to the delivery process must be also included when analyzing the response time of the processes running on the respective processor. We consider the delivery process during the schedulability analysis in the same way as the message transfer process.

6.5.1 STATIC SINGLE MESSAGE ALLOCATION (SM)

The first approach to scheduling of messages using TTP is to statically (off-line) schedule each of the messages into a slot of the TDMA cycle, corresponding to the node sending the message. This means that for each message we decide off-line to allocate space in one or more frames, space that can only be used by that particular message. In Figure 6.10 the frames are denoted by rectangles. In this particular example, it has been decided to allocate space for message m in slot S_1 of the first and third rounds.

Since the messages are dynamically produced by the processes, the exact moment a certain message is generated cannot be predicted. Hence, it can happen that certain frames will be left empty during execution. For example, if there is no message m in the *Out* queue (see Figure 3.6) when the slot S_1 of the first round in Figure 6.10 starts, that frame will carry no information. A message m produced immediately after slot S_1 starts to be transmitted, could then be carried by the frame scheduled in the slot S_1 of the third round.

In the SM approach, we consider that the slots can hold each at most one single message. This approach is well suited for application areas, like safety-critical automotive electronics, where the messages are typically short and the ability to easily diagnose the system (fewer messages in a frame are easier to observe) is critical. In the automotive electronics area messages are typically a couple of bytes, encoding signals like vehicle speed. However, for applications using larger messages, the SM approach leads to overheads due to the inefficient utilization of slot space when transmitting smaller size messages.

As each slot carries only one fixed, predetermined message, there is no interference among messages. If a message m misses its allocated frame it has to wait for the following slot assigned to m. The worst-case queuing delay W_m for a message m in this approach depends on the maximum time between consecutive slots of the same node carrying the message m. We denote this time by θ_m, illustrated in Figure 6.10, where we have a system

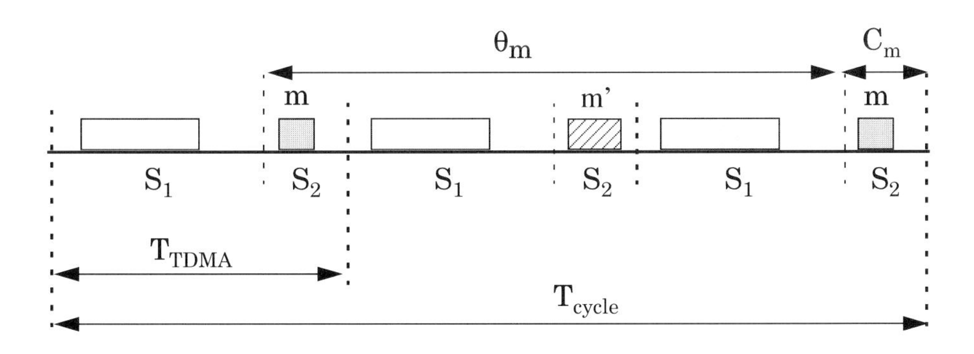

Figure 6.10: Worst-Case Arrival Time for SM

cycle of length T_{cycle}, consisting of three TDMA rounds. In this case, considering Equation (6.6), the worst-case response time r_m of a message m becomes:

$$r_m = \max_{q = 0, 1, 2...} (J_m + (q+1)\theta_m - qT_m + C_m). \qquad (6.15)$$

Therefore, the main aspect influencing schedulability of the messages is the way they are statically allocated to slots, which determines the values of θ_m. θ_m, as well as C_m, depend on the slot sizes which, in the case of SM, are determined by the size of the largest message sent from the corresponding node plus the bits for control and CRC, as imposed by the protocol.

As mentioned before, the analysis in [Tin94a], done for a simple TDMA protocol, assumes that $T_m \geq T_{TDMA}$. In the case of static message allocation with TTP (the SM and MM approaches), this translates to the condition $T_m \geq \theta_m$.

During the synthesis of the MEDL, the designer has to allocate the messages to slots in such a way that the process set is schedulable. Since the schedulability of the process set can be influenced by the synthesis of the MEDL only through the θ_m parameters, these are the parameters which have to be optimized.

Example 6.4: Let us consider the simple example depicted in Figure 6.11, where we have three processes, P_1, P_2, and P_3 running each on a different processor. When process P_1 finishes executing it sends message m_1 to process P_3 and message m_2 to process P_2. In the TDMA configurations presented in Figure 6.11, only the slot corresponding to the CPU running P_1 is important for our discussion and the other slots are represented with light gray. With the configuration in Figure 6.11a, where the message m_1 is allocated to the rounds one and four and the message m_2 is allocated to rounds two and three, process P_2 misses its deadline because of the release jitter due to the message m_2 in *Round 2*. How-

Figure 6.11: Optimizing the MEDL for SM and MM

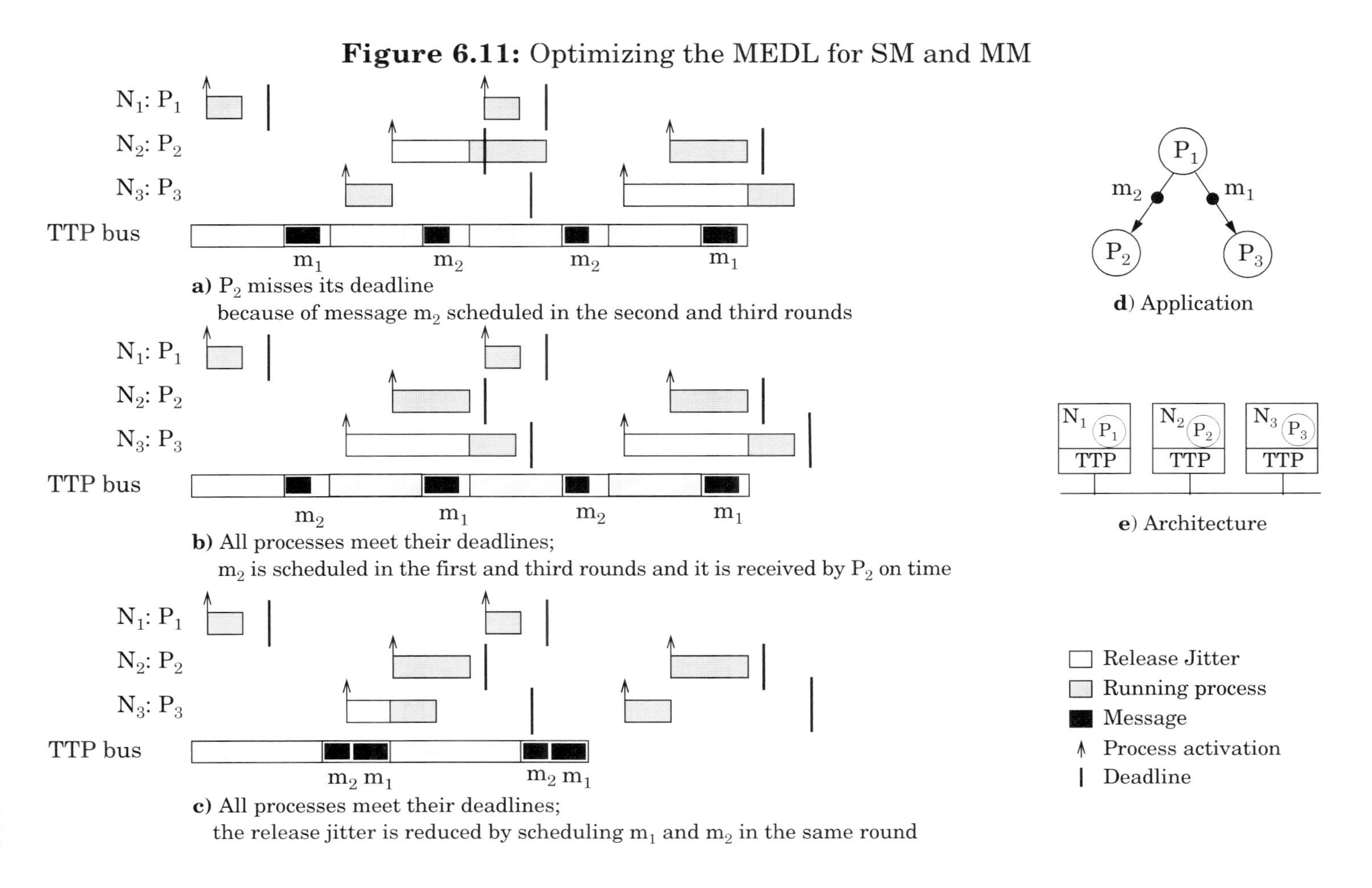

ever, if we have the TDMA configuration depicted in Figure 6.11b, where m_1 is allocated to the rounds two and four and m_2 is allocated to the rounds one and three, all the processes meet their deadlines.

■

6.5.2 STATIC MULTIPLE MESSAGE ALLOCATION (MM)

This second approach is an extension of the first one. In this approach we allow more than one message to be statically assigned to a slot and all the messages transmitted in the same slot are packed together in a frame. As with the SM approach, there is no interference among messages, so the worst case access delay for a message m is the maximum time between consecutive slots of the same node carrying the message m, θ_m. It is also assumed that $T_m \geq \theta_m$.

However, this approach offers more freedom during the synthesis of the MEDL. We have now to decide also on how many and which messages should be packed in a slot. This allows more flexibility in optimizing the θ_m parameter.

Example 6.5: To illustrate this, let us consider the same example depicted in Figure 6.11. With the MM approach, the TDMA configuration can be arranged as depicted in Figure 6.11c, where the messages m_1 and m_2 are put together in the same slot in the first and second rounds. Thus, the deadline is met and the release jitter is further reduced compared to the case presented in Figure 6.11b where process P_3 was experiencing a large release jitter.

■

6.5.3 DYNAMIC MESSAGE ALLOCATION (DM)

The previous two approaches have statically allocated one or more messages to their corresponding slots. This third approach considers that the messages are dynamically allocated to frames, as they are produced.

Thus, as soon as a message is produced, it is placed in the *Out* queue (see Figure 3.6 on page 56), ordered according to the message priorities. When the transfer process is activated on node N_i, it removes from the head of the queue a number of messages, such that the total size does not exceed the length of the data field of the frame allocated to slot S_i. Since the messages are sent dynamically, we have to identify them in a certain way so that they are recognized when the frame arrives at the delivery process. We consider that each message has several identifier bits appended at the beginning of the message.

We dynamically pack messages into frames in the order they are sorted in the queue, thus, the worst-case queuing delay to the communication channel for a message m depends on the number of messages queued ahead of it.

The analysis in [Tin94a] for a simple TDMA bus, presented in Section 6.4.2, bounds the number of queued-ahead *packets* of messages of higher priority than message m, as in their case it is considered that a message can be split into packets before it is transmitted on the communication channel (Equation 6.13). We use the same analysis but we have to apply it for the number of *messages* instead of packets. We have to consider that messages can be of different sizes as opposed to packets which always are of the same size.

Therefore, the total *size* of higher priority messages queued ahead of a message m, in the worst case, is:

$$I_m(w_m(q)) = \sum_{\forall m_j \in hp(m)} \left\lceil \frac{w_m(q) + r_{S(m_j)}}{T_j} \right\rceil S_j , \qquad (6.16)$$

where S_j is the size of the message m_j, $r_{S(j)}$ is the response time of the process sending message m_j, and T_j is the period of the message m_j.

Further, we calculate the worst-case time that a message m spends in the *Out* queue. The number of TDMA rounds needed, in the worst case, for a message m placed in the queue to be removed from the queue for transmission is

$$\left\lceil \frac{(q+1)S_m + I_m(w_m(q))}{S_s} \right\rceil, \tag{6.17}$$

where S_m is the size of the message m and S_s is the size of the slot transmitting m (we assume, in the case of DM, that for any message x, $S_x \leq S_s$). This means that the worst case time a message m spends in the *Out* queue is given by

$$w_m(q) = \left\lceil \frac{(q+1)S_m + I_m(w_m(q))}{S_s} \right\rceil T_{TDMA}, \tag{6.18}$$

where T_{TDMA} is the time taken for a TDMA round.

Since the size of the messages is fixed for a given application, the parameter that will be optimized during the synthesis of the MEDL is the slot size.

Example 6.6: To illustrate how the slot size influences schedulability, let us consider the example in Figure 6.12 where we have the same setting as for the example in Figure 6.11. The difference is that we consider message m_1 having a higher priority than message m_2 and we schedule the messages dynamically as they are produced.

With the configuration in Figure 6.12a message m_1 will be dynamically scheduled first in the slot of the first round, while message m_2 will wait in the *Out* queue until the next round comes, thus causing process P_2 to miss its deadline. However, if we enlarge the slot so that it can accommodate both messages, message m_2 does not have to wait in the queue and it is transmitted in the same slot as m_1. Therefore, P_2 will meet its deadline as presented in Figure 6.12b.

However, in general, increasing the length of slots does not necessarily improve schedulability, as it delays the communication of messages generated by other nodes.

■

6.5.4 DYNAMIC PACKET ALLOCATION (DP)

This approach is an extension of the previous one, as we allow the messages to be split into packets before they are transmitted on the communication channel. We consider that each slot has a size that accommodates a frame with the data field being a multiple of the packet size. The analysis for this case is similar to the one outlined in Section 6.4.2 for the simple TDMA bus.

This approach is well suited for the application areas that typically have large message sizes. By splitting messages into packets, we can obtain a higher utilization of the bus and reduce the release jitter. However, since each packet has to be identified as belonging to a message, and messages have to be split at the sender and reconstructed at the destination, the overhead becomes higher than in the previous approaches.

In the previous approach (DM) the optimization parameter for the synthesis of the MEDL was the size of the slots. With this approach we can also decide on the packet size, which becomes another optimization parameter.

Example 6.7: Consider the example in Figure 6.12c where messages m_1 and m_2 have a size of 6 bytes each. The packet size is considered to be 4 bytes and the slot corresponding to the messages has a size of 12 bytes (three packets) in the TDMA configuration. Since message m_1 has a higher priority than m_2, it will be dynamically scheduled first in the slot of the first round and it will need two packets. In the third packet the first 4 bytes of m_2 are placed. Thus, the remaining 2 bytes of message m_2 have to wait for the next round, causing process P_2 to miss its deadline. However, if we change the packet size to 3 bytes and keep the same size of 12 bytes for

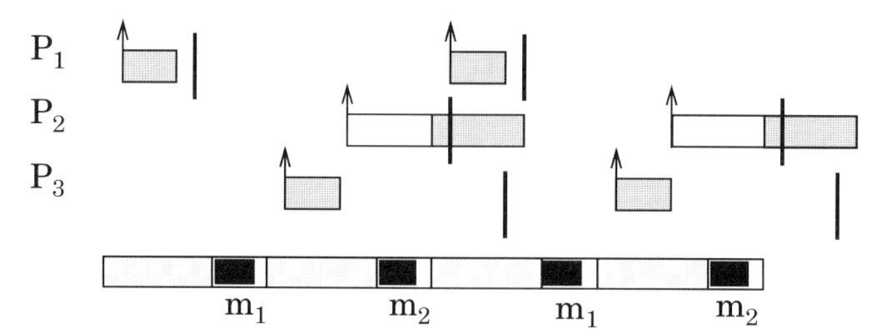

a) P_2 misses its deadline; there is no space in the slot of the first round to schedule the lower priority message m_2

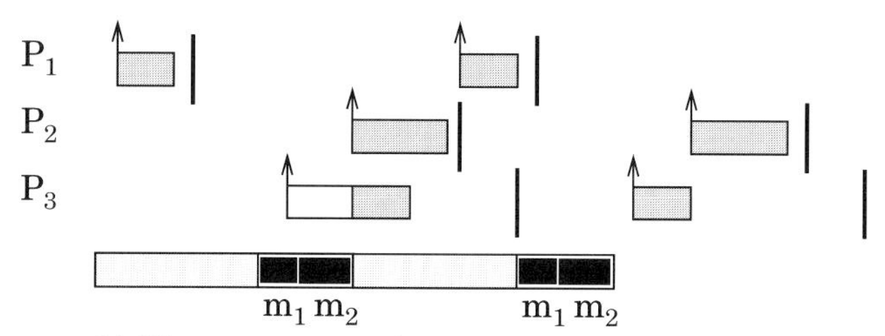

b) All processes meet their deadlines; the slot has been enlarged to hold both messages

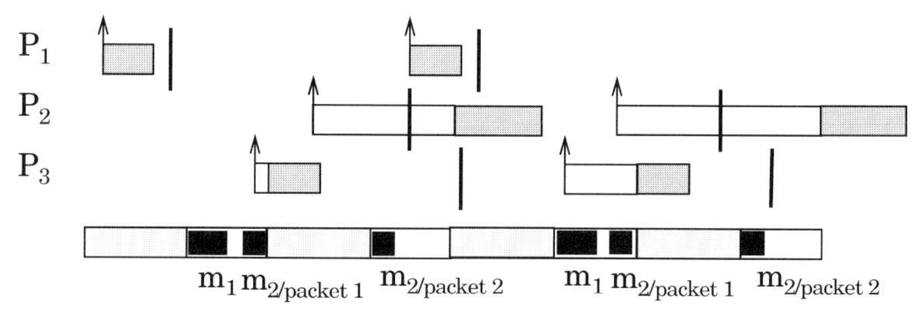

c) P_2 misses its deadline; the slot is too small to hold both packets of message m_2

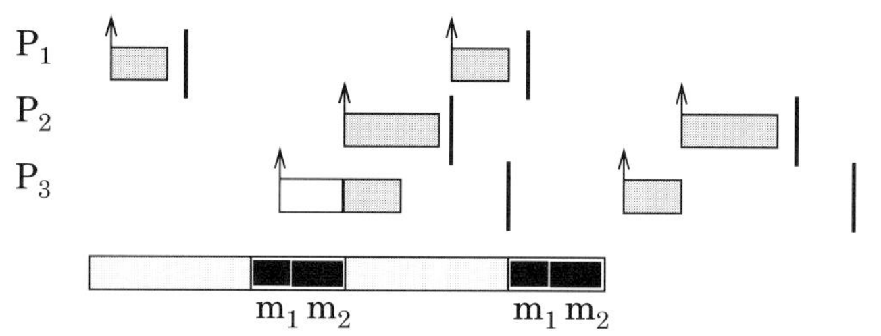

d) All processes meet their deadlines; the slot has been enlarged to hold 4 packets instead of 3

☐ Release Jitter ☐ Running process ■ Message
↑ Process activation | Deadline

Figure 6.12: Optimizing the MEDL for DM and DP

the slot, we have four packets in the slot corresponding to the CPU running P_1 (Figure 6.12d). Message m_1 will be dynamically scheduled first and will need 2 packets in the slot of the first round. Hence, m_2 can be sent in the same round so that P_2 can meet its deadline.

■

In the above example, with one single sender processor and the particular message and slot sizes as given, the problem seems to be simple. This is, however, not the case in general. For example, the packet size which fits a particular node can be unsuitable in the context of the messages and slot size corresponding to another node. At the same time, reducing the packets size increases the overheads due to the transfer and delivery processes.

6.6 Schedulability Analysis for Event-Driven Systems

The previous sections have built, step by step, a response time analysis for event driven systems that is able to handle:

- data dependencies in applications,
- distributed architectures,
- the details of several communication protocols, and
- control dependencies.

Therefore, at this point we have two analyses (see the boxes with a thick border in Figure 6.2):

1. A response time analysis for applications with data and control dependencies distributed over a multi-processor architecture that uses TTP as the communication protocol. In Section 6.7 we will show how this analysis can be used to drive a bus access optimization process for the time-triggered protocol.
2. A similar response time analysis, but for CAN. This analysis

will be used in Chapter 8 for the event-triggered cluster of a multi-cluster system.

6.6.1 DEGREE OF SCHEDULABILITY

To determine if an application is schedulable, it is enough to compare the response times with the deadlines. However, in order to drive our optimization heuristics aiming at obtaining a schedulable system at a low cost, we need a metric that is able to indicate which of the design alternatives is "more schedulable" than the others.

Thus, in order to guide the optimization process, we have used as a cost function the "degree of schedulability." Our cost function is similar to that in [Tin92] in the case an application is not schedulable (c_1). However, in order to distinguish between several schedulable applications, we have introduced the second expression c_2, which measures, for a feasible design alternative, the total difference between the response times and the deadlines. We use the following function in order to express the degree of schedulability:

$$\textit{degree of schedulability}(\textit{design}) = \begin{cases} c_1 = \sum_{i=1}^{n} max(0, R_i - D_i), \text{ if } c_1 > 0 \\ c_2 = \sum_{i=1}^{n} (R_i - D_i), \text{ if } c_1 = 0 \end{cases}$$

where n is the number of processes in the application, R_i is the worst-case response time of a process P_i, and D_i is the deadline of a process P_i. If the application is not schedulable, there exists at least one R_i greater than the deadline D_i, therefore the term c_1 of the function will be positive. In this case the cost function is equal to c_1.

However, if the application is schedulable, then each R_i is smaller than the corresponding deadline D_i. In this case $c_1 = 0$ and we use c_2 as the cost function, as it is able to differentiate between two alternatives, both leading to a schedulable application. For a given design implementation leading to a schedulable application, a smaller c_2 means that we have improved the response times of the processes, so the application can be potentially implemented on a cheaper hardware architecture (with slower processors and/or bus, but without increasing the number of processors or buses).

6.7 Bus Access Optimization

Once a schedulability analysis for event-driven distributed real-time systems is in place, our problem is to analyze the schedulability of a given process set and to synthesize the MEDL of the TTP controllers (and consequently the MHTTs) so that the application is schedulable on an as cheap as possible architecture. The optimization is performed on the parameters which have been identified for each of the four approaches to message scheduling discussed before (see Section 6.5). In order to guide the optimization process, we have used as a cost function the degree of schedulability calculated for a given MEDL implementation.

For a given application, we are interested to synthesize a MEDL such that the degree of schedulability cost function is minimized. We are also interested to evaluate in different contexts the four approaches to message scheduling, thus offering the designer a decision support for choosing the approach that best fits his application.

The MEDL synthesis problem belongs to the class of exponential complexity problems, therefore we are interested to develop heuristics that are able to find accurate results in a reasonable time. We have developed optimization algorithms corresponding to each of the four approaches to message scheduling. A first set of algorithms presented in Section 6.7.1 is based on simple and fast greedy heuristics. In Section 6.7.2 we introduce a second class of heuristics which aims at finding near-optimal solutions using the simulated annealing (SA) algorithm.

6.7.1 GREEDY HEURISTICS

We have developed greedy heuristics for each of the four approaches to message scheduling. The main idea of the heuristics is to minimize the cost function by incrementally trying to reduce the communication delay of messages and, by this, the release jitter of the processes.

The only way to reduce the release jitter in the SM and MM approaches is through the optimization of the θ_m parameters. This is achieved by a proper placement of messages into slots (see Figure 6.11).

The OptimizeSM algorithm presented in Figure 6.13 starts by deciding on a size ($size_{S_i}$) for each of the slots. The slot sizes are set to the minimum size that can accommodate the largest message sent by the corresponding node (lines 1–4 in Figure 6.13). In this approach a slot can carry at most one message, thus slot sizes larger than this size would lead to larger response times.

Then, the algorithm has to decide on the number of rounds, thus determining the size of the MEDL. Since the size of the MEDL is physically limited, there is a limit to the number of rounds (e.g., 2, 4, 8, 16 depending on the particular TTP controller implementation). However, there is a minimum number of rounds MinRounds, which are necessary for a certain application, and depends on the number of messages transmitted (lines 5–9). For example, if the processes mapped on node N_1 send in total seven messages then we have to decide on at least seven rounds in order to accommodate all of them (in the SM approach there is at most one message per slot). Several numbers of rounds RoundsNo are tried out by the algorithm starting from MinRounds up to MaxRounds (lines 15–31).

For a given number of rounds (that determine the size of the MEDL), the initially empty MEDL has to be populated with messages in such a way that the cost function is minimized. In order to apply the schedulability analysis, which is the basis for the cost function, a *complete* MEDL has to be provided. A complete MEDL contains at least one instance of every message that has to be transmitted between the processes on different processors. A

```
OptimizeSM
1  -- set the slot sizes
2  for each node N_i do
3      size_{S_i} = max(size of messages m_j sent by node N_i)
4  end for
5  -- find the minimum number of rounds that can hold all the messages
6  for each node N_i do
7      nm_i = number of messages sent from N_i
8  end for
9  MinRounds = max (nm_i)
10 -- create a minimal complete MEDL
11 for each message m_i do
12     find round in [1..MinRounds] that has an empty slot for m_i
13     place m_i into its slot in round
14 end for
15 for each RoundsNo in [MinRounds...MaxRounds] do
16     -- insert messages in such a way that the cost is minimized
17     repeat
18         for each process P_i that receives a message m_i do
19             if D_i – R_i is the smallest so far then m = m_{P_i} end if
20         end for
21         for each round in [1..RoundsNo] do
22             place m into its corresponding slot in round
23             calculate the CostFunction
24             if the CostFunction is smallest so far then
25                 BestRound = round
26             end if
27             remove m from its slot in round
28         end for
29         place m into its slot in BestRound if one was identified
30     until the CostFunction is not improved
31 end for
end OptimizeSM
```

Figure 6.13: Greedy Heuristic for SM

minimal complete MEDL is constructed from an empty MEDL by placing one instance of every message m_i into its corresponding empty slot of a round (lines 10–14).

Example 6.8: In Figure 6.11a, for example, we have a MEDL composed of four rounds. We get a minimal complete MEDL, for example, by assigning m_2 and m_1 to the slots in rounds three and four, and letting the slots in rounds one and two empty.

■

However, such a MEDL might not lead to a schedulable system. The degree of schedulability can be improved by inserting instances of messages into the available places in the MEDL, thus minimizing the θ_m parameters.

Example 6.9: For example, in Figure 6.11a, by inserting another instance of the message m_1 in the first round and m_2 in the second round leads to P_2 missing its deadline, while in Figure 6.11b inserting m_1 into the second round and m_2 into the first round leads to a schedulable system.

■

Our algorithm repeatedly adds a new instance of a message to the current MEDL in the hope that the cost function will be improved (lines 16–30). In order to decide an instance of which message should be added to the current MEDL, a simple heuristic is used. We identify the process P_i which has the most "critical" situation, meaning that the difference between its deadline and response time, $D_i - R_i$, is minimal compared with all other processes. The message to be added to the MEDL is the message $m = m_{P_i}$ received by the process P_i (lines 18–20). Message m will be placed into that round (BestRound) which corresponds to the smallest value of the cost function (lines 21–28). The algorithm stops if the cost function cannot be further improved by adding more messages to the MEDL.

The OptimizeMM algorithm is similar to OptimizeSM. The main difference is that in the MM approach several messages can be placed into a slot (which also decides its size), while in the SM approach there can be at most one message per slot. Also, in the case of MM, we have to take additional care that the slots do not exceed the maximum allowed size for a slot.

The situation is simpler for the dynamic approaches, namely DM and DP, since we only have to decide on the slot sizes and, in the case of DP, on the packet size. For these two approaches, the placement of messages is dynamic and has no influence on the cost function. The OptimizeDM algorithm in see Figure 6.14 starts with the first slot $S_i = S_1$ of the TDMA round and tries to find that size ($BestSize_{S_i}$) which corresponds to the smallest CostFunction (lines 4–14 in Figure 6.14). This slot size has to be large enough ($S_i \geq MinSize_{S_i}$) to hold the largest message to be transmitted in this slot, and within bounds determined by the particular TTP controller implementation (e.g., from 2 bits up to MaxSize = 32 bytes). Once the size of the first slot has been determined, the algorithm continues in the same manner with the next slots (lines 7–12).

```
OptimizeDM
1   for each node N_i do
2       MinSize_{S_i} = max(size of messages m_j sent by node N_i)
3   end for
4   -- identifies the size that minimizes the cost function
5   for each slot S_i do
6       BestSize_{S_i} = MinSize_{S_i}
7       for each SlotSize in [MinSize_{S_i}...MaxSize] do
8           calculate the CostFunction
9           if the CostFunction is best so far then
10              BestSize_{S_i} = SlotSize_{S_i}
11          end if
12      end for
13      size_{S_i} = BestSize_{S_i}
14  end for
end OptimizeDM
```

Figure 6.14: Greedy Heuristic for DM

The OptimizeDP algorithm has also to determine the proper packet size. This is done by trying all the possible packet sizes given the particular TTP controller. For example, it can start from 2 bits and increment with the "smallest data unit" (typically two bits) up to 32 bytes. In the case of the OptimizeDP algorithm the slot size has to be determined as a multiple of the packet size and within certain bounds depending on the TTP controller.

6.7.2 SIMULATED ANNEALING STRATEGY

We have also developed an optimization procedure based on a simulated annealing (SA) strategy, described in Appendix A. The main characteristic of such a strategy is that it tries to find the global optimum by randomly selecting a new solution from the neighbors of the current solution.

The neighbors of the current solution are obtained depending on the chosen message scheduling approach. For SM, the next solution is obtained from the current one by inserting or removing a message in one of its corresponding slots. In the case of MM, we have to take additional care that the slots do not exceed the maximum allowed size (which depends on the controller implementation), as we can allocate several messages to a slot. For these two static approaches we also decide on the number of rounds in a cycle (e.g., 2, 4, 8, 16, limited by the size of the memory implementing the MEDL). In the case of DM, the neighboring solution is obtained by increasing or decreasing the slot size within the bounds allowed by the particular TTP controller implementation, while in the DP approach we also increase or decrease the packet size.

We have also tuned the parameters *TI* (initial temperature), *TL* (temperature length), and ε (cooling ratio) that define the cooling schedule (see Appendix A for the details on these parameters). For example, for the graphs with 320 nodes, *TI* is 300, *TL* is 500 and ε is 0.95. The algorithm stops if for three consecutive temperatures no new solution has been accepted.

6.8 Experimental Evaluation

In Section 6.8.1, the results for the schedulability analysis of systems with data and control dependencies (Section 6.3) are presented. Section 6.8.2 first presents the experimental results for the schedulability analysis with the TTP (Section 6.4), comparing the four message scheduling approaches. Then, we present the results obtained for the communication synthesis problem outlined in Section 6.7. The approaches presented in this chapter are further evaluated in Section 6.8.3 using the vehicle cruise controller model from Section 2.3.3.

6.8.1 SCHEDULABILITY ANALYSIS FOR SYSTEMS WITH CONTROL AND DATA DEPENDENCIES

In this section we present some experimental results regarding the schedulability analysis for conditional process graphs which has been discussed in Section 6.3. The two main aspects we were interested in are the quality of the schedulability analysis and the scalability of the algorithms for large examples. A better quality of a schedulability analysis, in this case, means a lower degree of pessimism. A set of massive experiments were performed on conditional process graphs generated for experimental purpose.

We considered architectures consisting of 2, 4, 6, 8 and 10 processors. Forty processes were assigned to each node, resulting in applications of 80, 160, 240, 320 and 400 processes, having 2, 4, 6, 8 and 10 conditions, respectively. The number of unconditional subgraphs varied for each application dimension depending on the number of conditions and the randomly generated structure of the CPGs. For example, for applications with 400 processes, the maximum number of unconditional subgraphs was 64.

Thirty applications were generated for each dimension, thus a total of 150 graphs were used for experimental evaluation. Worst case execution times were assigned randomly using both uniform and exponential distribution. The experiments were also run on a SUN Ultra 10 workstation.

In order to compare the quality of the schedulability approaches, we need a cost function that captures, for a certain system, the difference in quality between the schedulability approaches proposed (Section 6.3). Our cost function is the difference between the deadline and the worst-case delay of a CPG, summed for all the CPGs in the system:

$$\textit{cost function} = \sum_{i=1}^{n} (D_{G_i} - \delta_{G_i}) \tag{6.19}$$

where n is the number of CPGs in the application, δ_{G_i} is the worst-case delay of the CPG G_i, and D_{G_i} is the deadline on G_i. A higher value for this cost function, for a given system, means that the corresponding approach produces better results (schedulability analysis is less pessimistic).

For each of the 150 generated example systems and each of the five schedulability analyses we have calculated the cost function mentioned previously, based on results produced with the algorithms described in Section 6.3. These values, for a given system, differ from one analysis to another, with the BF being the least pessimistic approach and therefore having the largest value for the cost function.

We are interested to compare the approaches based on the values obtained for the cost function (Equation 6.19). Figure 6.15a presents the average percentage deviations of the cost function obtained in each of the approaches, compared to the value of the cost function obtained with the BF approach. A smaller value for the percentage deviation means a larger cost function, thus a

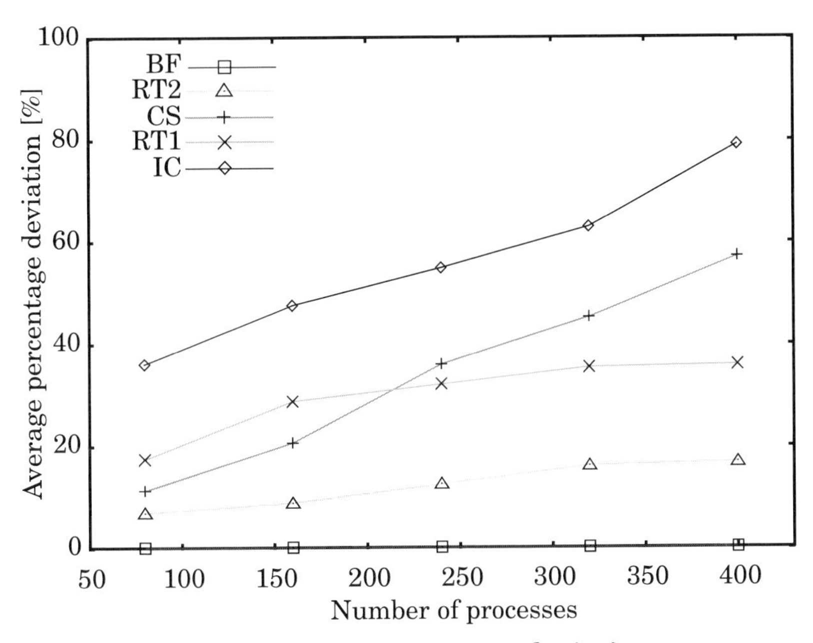

a) Average percentage deviations

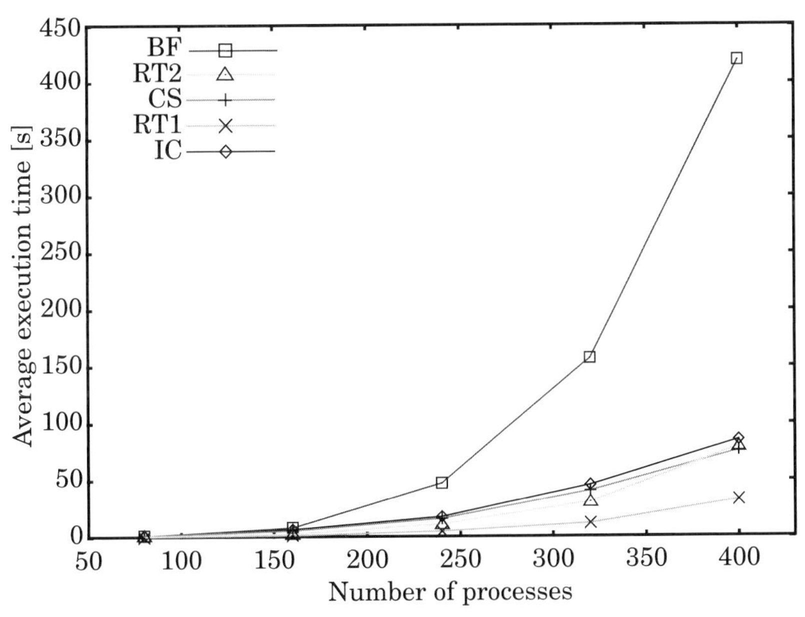

b) Average execution times

Figure 6.15: Comparison of the Schedulability Approaches for CPGs

better result. The percentage deviation is calculated according to the formula:

$$\text{deviation} = \frac{\text{cost}_{approach} - \text{cost}_{best}}{\text{cost}_{best}} \cdot 100 \qquad (6.20)$$

Figure 6.15b presents the average runtime of the algorithms, in seconds.

The brute force approach, BF, performs best in terms of quality and obtains the largest values for the cost function at the expense of a large execution time. The execution time can be up to 7 minutes for large graphs of 400 processes, 10 conditions, and 64 unconditional subgraphs. At the other end, the straightforward approach IC that ignores the conditions, performs worst and becomes more and more pessimistic as the system size increases. As can be seen from Figure 6.15, IC has even for smaller systems of 160 processes (3 conditions, maximum 8 unconditional subgraphs) a 50% worse quality than the brute force approach, with almost 80% loss in quality, in average, for large systems of 400 processes. It is interesting to mention that the low quality IC approach has also an average execution time which is equal or comparable to the much better quality heuristics (except BF, of course). This is because it tries to improve on the worst-case delays through the iterative loop presented in DelayEstimate, Figure 6.3.

Let us turn our attention to the three approaches CS, RT1, and RT2 that, like BF, consider conditions during the analysis but also try to perform a trade-off between quality and execution time. Figure 6.15 shows that the pessimism of the analysis is dramatically reduced by considering the conditions during the analysis. The RT1 and RT2 approaches, which visit each unconditional subgraph, perform in average better than the CS approach that considers condition separation for the whole graph. However, CS is comparable in quality with RT1, and even performs better for graphs of size smaller than 240 processes (4 conditions, maximum 16 subgraphs).

The RT2 analysis that tries to improve the worst case response times using the MaxSeparations, as opposed to RT1, performs best among the non-brute-force approaches. As can be seen from Figure 6.15, RT2 has less than 20% average deviation from the solutions obtained with the brute force approach. However, if faster runtimes are needed, RT1 can be used instead, as it is twice faster in execution time than RT2.

We were also interested to compare the approaches with respect to the number of unconditional subgraphs and the number of conditional process graphs in an application. For the results depicted in Figure 6.16 we have assumed CPGs consisting of 2, 4, 8, 16, and 32 unconditional subgraphs of maximum 50 processes each, allocated to 8 processors. Figure 6.16 shows that as the number of subgraphs increases, the differences between the approaches grow while the ranking among them remains the same, as resulted from Figure 6.15. The CS approach performs better than RT1 with a smaller number of subgraphs, but RT1 becomes better as the number of subgraphs in the CPGs increases.

Figure 6.17 presents on a logarithmic scale the average percentage deviations for systems consisting of 1, 2, 3, 4 and 5 conditional process graphs of 160 nodes each. As the number of conditional process graphs increases, the IC and CS approaches become more pessimistic. However, RT1 and RT2 perform very well, with RT2 being the least pessimistic approach (except the BF approach, which is not depicted in Figure 6.17).

6.8.2 SCHEDULABILITY ANALYSIS WITH TTP AND BUS ACCESS OPTIMIZATION

For the evaluation of our message scheduling approaches over TTP we used applications generated for experimental purpose. We considered architectures consisting of 2, 4, 6, 8 and 10 nodes.

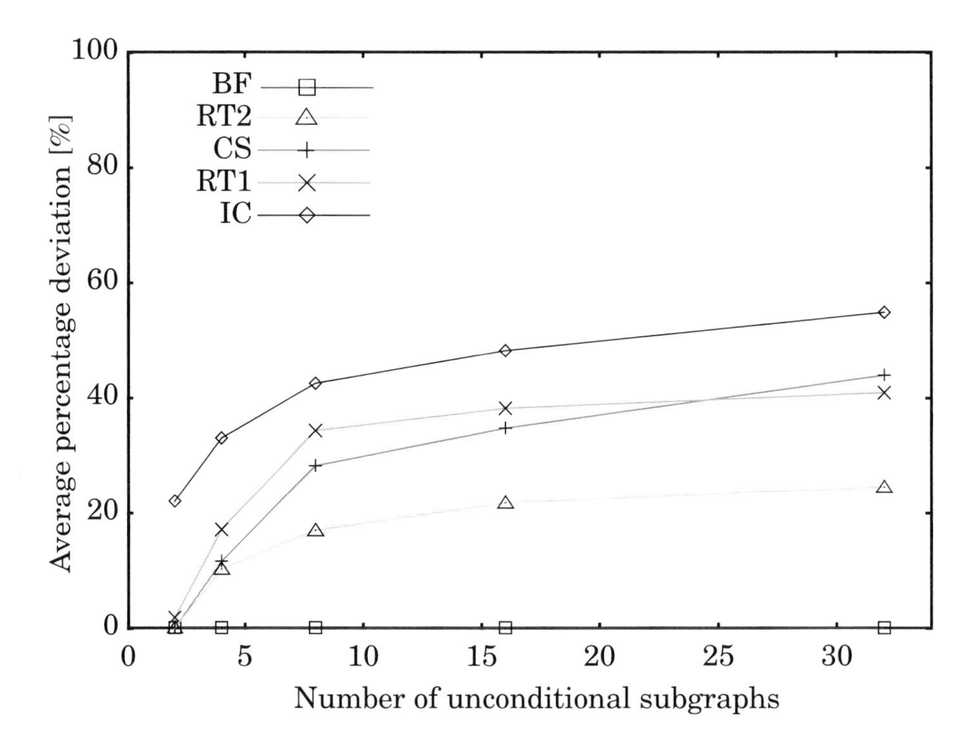

Figure 6.16: Comparison of the Schedulability Analysis Approaches for CPGs Based on the Number of Unconditional Subgraphs

Forty processes were assigned to each node, resulting in sets of 80, 160, 240, 320 and 400 processes. Thirty applications were generated for each of the five dimensions. Thus, a total of 150 applications were used for experimental evaluation. Worst-case computation times, periods, deadlines, and message lengths were assigned randomly within certain intervals. For the communication channel we considered a transmission speed of 256 Kbps. The maximum length of the data field in a slot was 32 bytes and the frequency of the TTP controller was chosen to be 20 MHz. These experiments were also run on a SUN Ultra 10 workstation.

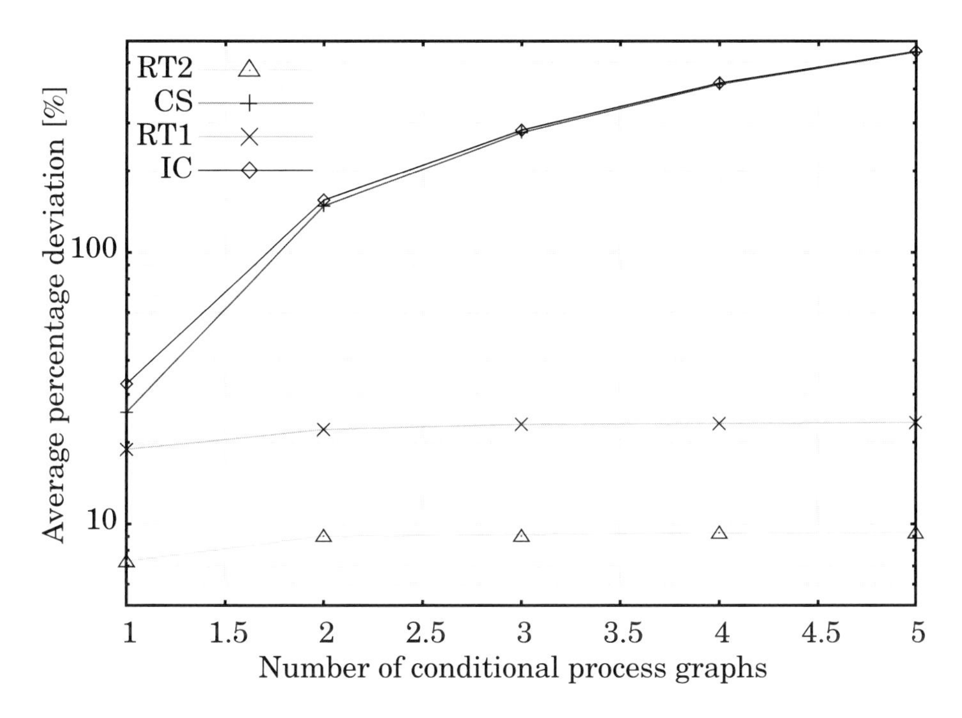

Figure 6.17: Comparison of the Schedulability Analysis Approaches Based on the Number of CPGs

For each of the 150 generated examples and each of the four message scheduling approaches we have obtained the near-optimal values for the cost function (degree of schedulability, Section 6.6.1) as produced by our SA based algorithm (see Section 6.7.2). For a given example, these values might differ from one message passing approach to another, as they depend on the optimization parameters and the schedulability analysis which are particular for each approach. Figure 6.18 presents a comparison based on the average percentage deviation of the cost function obtained for each of the four approaches, from the minimal value among them. The percentage deviation is calculated according to the Equation 6.20.

The DP approach is, generally, able to achieve the highest degree of schedulability, which in Figure 6.18 corresponds to the smallest deviation. In the case the packet size is properly selected, by scheduling messages dynamically we are able to efficiently use the available space in the slots, and thus reduce the release jitter. However, by using the MM approach we can obtain almost the same result if the messages are carefully allocated to slots as does our optimization strategy.

Moreover, in the case of larger process sets, the static approaches suffer significantly less overhead than the dynamic approaches. In the SM and MM approaches the messages are uniquely identified by their position in the MEDL. However, for the dynamic approaches we have to somehow identify the dynamically transmitted messages and packets. Hence, for the DM approach we consider that each message has several identi-

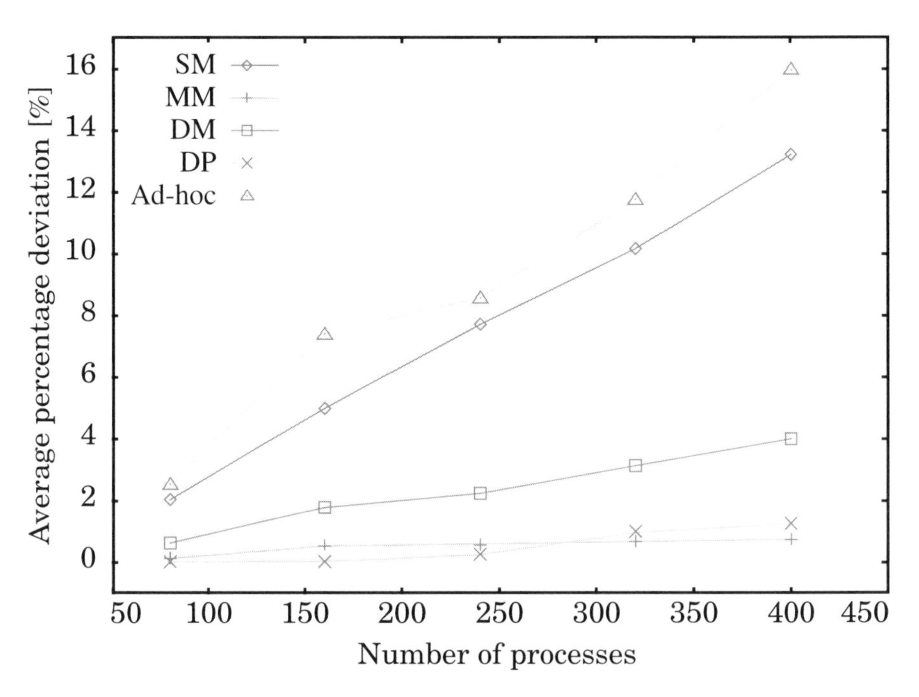

Figure 6.18: Comparison of the Four Approaches to Message Scheduling over TTP

fier bits appended at the beginning of the message, while for the DP approach the identification bits are appended to each packet. Not only the identifier bits add to the overhead, but in the DP approach, the transfer and delivery processes (see Figure 3.6 on page 56) have to be activated for each sending and receiving of a packet, and so interfere with the other processes.

Therefore, for larger applications (e.g., 400 processes), MM outperforms DP, as DP suffers from large overhead due to its dynamic nature. DM performs worse than DP because it does not split the messages into packets, and this results in a mismatch between the size of the messages dynamically queued and the slot size, leading to unused slot space that increases the jitter. SM performs the worst as it does not permit much room for improvement, leading to large amounts of unused slot space. Also, DP has produced a MEDL that resulted in schedulable applications for 1.33 times more cases than the MM and DM. MM, in its turn, produced two times more schedulable results than the SM approach.

Together with the four approaches to message scheduling a so called ad-hoc approach is also presented. The ad-hoc approach performs scheduling of messages without trying to optimize the access to the communication channel. The ad-hoc solutions are based on the MM approach and consider a design with the TDMA configuration consisting of a simple, straightforward allocation of messages to slots. The lengths of the slots were selected to accommodate the largest message sent from the respective node. Figure 6.18 shows that the ad-hoc alternative is constantly outperformed by any of the optimized solutions. This demonstrates that significant gains can be obtained by optimization of the parameters defining the access to the communication channel.

Next, we have compared the four approaches with respect to the number of messages exchanged between different nodes and the maximum message size allowed. For the results depicted in figures 6.19 and 6.20 we have assumed applications of 80 pro-

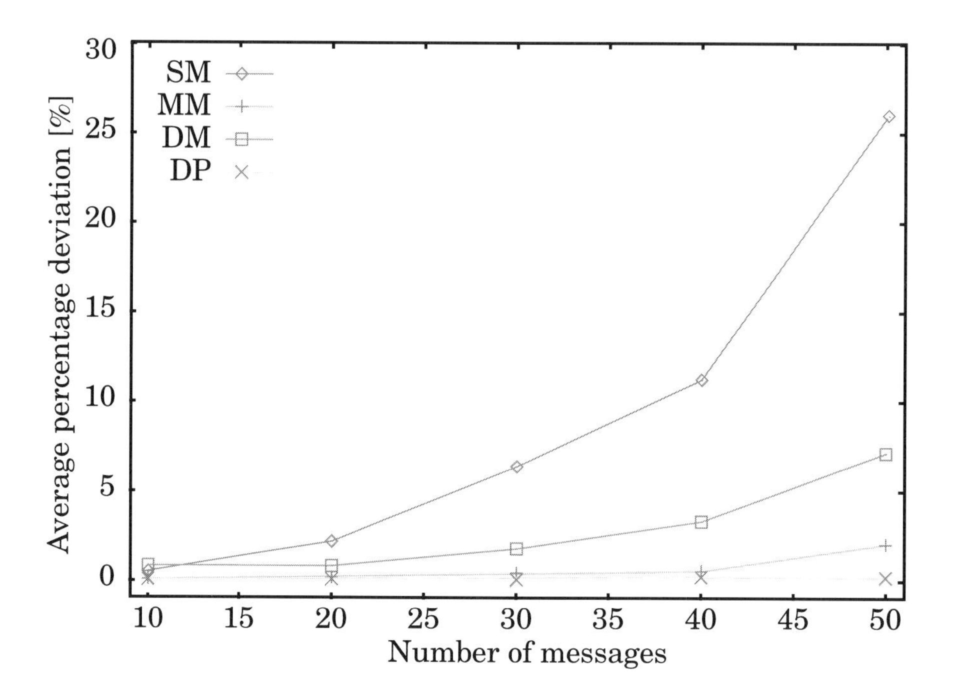

Figure 6.19: Four Approaches to Message Scheduling over TTP: The Influence of the Messages Number

cesses allocated to 4 nodes. Figure 6.19 shows that, as the number of messages increases, the difference between the approaches grows while the ranking among them remains the same. The same holds for the case when we increase the maximum allowed message size (Figure 6.20), with a notable exception: for large message sizes MM becomes better than DP, since DP suffers from larger overhead due to its dynamic nature.

The above comparison between the four message scheduling alternatives is mainly based on the issue of schedulability. However, when choosing among the different policies, several other parameters can be of importance. For example, a static allocation of messages can be beneficial from the point of view of test-

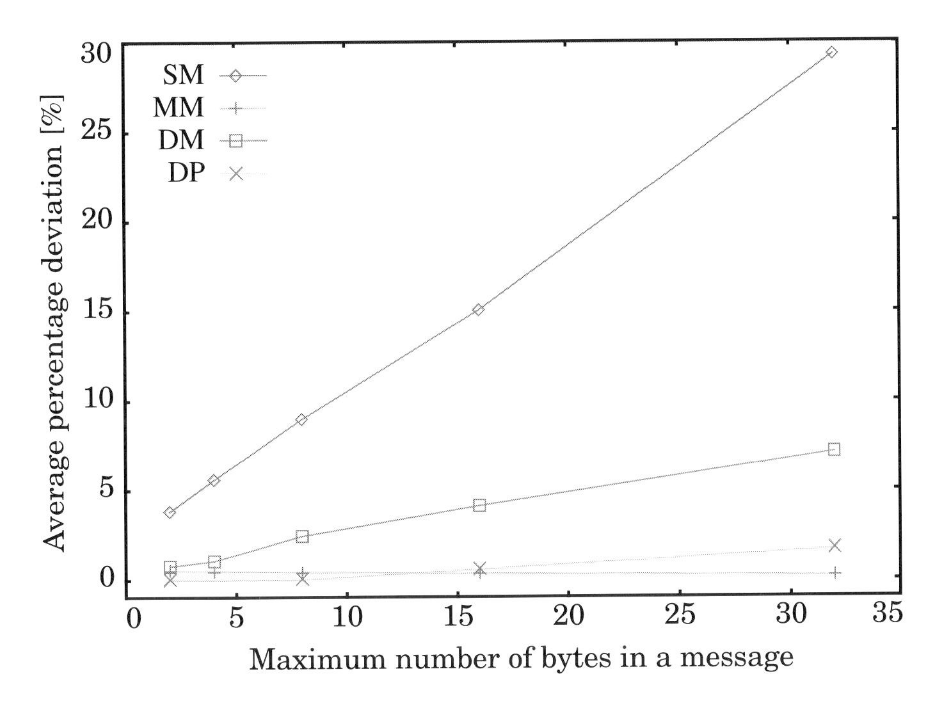

Figure 6.20: Four Approaches to Message Scheduling over TTP: The Influence of the Message Sizes

ing and debugging and has the advantage of simplicity. Similar considerations can lead to the decision not to split messages. In any case, however, optimization of the bus access scheme is highly desirable.

We were also interested in the quality of our greedy heuristics. Thus, we have run all the examples presented above, using the greedy heuristics and compared the results with those produced by the SA based algorithm. Table 6.2 shows the average and maximum percentage deviations of the cost function values produced by the greedy heuristics from those generated with SA, for each of the application dimensions. All the four greedy heuristics perform very well, with less than 2% loss in quality compared to the results produced by the SA algorithms. The execution times for

Table 6.2: Percentage Deviations for the Greedy Heuristics Compared to Simulated Annealing

Processes		80	160	240	320	400
SM	aver.	0.12%	0.19%	0.50%	1.06%	1.63%
	max.	0.81%	2.28%	8.31%	31.05%	18.00%
MM	aver.	0.05%	0.04%	0.08%	0.23%	0.36%
	max.	0.23%	0.55%	1.03%	8.15%	6.63%
DM	aver.	0.02%	0.03%	0.05%	0.06%	0.07%
	max.	0.05%	0.22%	0.81%	1.67%	1.01%
DP	aver.	0.01%	0.01%	0.05%	0.04%	0.03%
	max.	0.05%	0.13%	0.61%	1.42%	0.54%

the greedy heuristics were more than two orders of magnitude smaller than those with SA.

6.8.3 The Vehicle Cruise Controller

We have applied our approaches in sections 6.3 and 6.4 to the real-life example implementing a vehicle cruise controller described in Section 2.3.3:

- The hardware architecture considered consists of five nodes interconnected by a TTP bus, and is presented in Figure 2.7a on page 37.
- We have used the software architecture for event-driven systems, outlined in Section 3.4.
- We have applied the analyses in Section 6.3 to the mapped CPG, modeling the CC system presented in Figure 2.9 on page 40, but without considering the messages (depicted with solid circles in that figure). The deadline in this case has been set to 130 ms.

- We have also evaluated the scheduling approaches presented in Section 6.4 using the CC model in Figure 2.9, considering in this case a deadline of 500 ms.

For the approaches in Section 6.3, that aim at reducing the pessimism of the analysis by using the conditions in the model, we have obtained the following results. Without considering the conditions, IC obtained a worst case delay of 138 ms, thus the system resulted as being unschedulable. The system has also been declared as unschedulable by the Conditions Separation (CS) approach, which has produced a result of 132 ms.

However, the Brute Force approach (BF) has produced a worst-case delay of 124 ms which proves that the system implementing the vehicle cruise controller is, in fact, schedulable. Both Relaxed Tightness alternatives (RT1 and RT2) have produced the same worst case delay of 124 ms as the BF.

For the techniques in Section 6.4, where we have proposed four message scheduling approaches using the TTP, we have the following results concerning the cruise controller example. The ad-hoc solution and the SM approach failed to produce a schedulable solution (in both cases, 27 out of 32 processes had a response time larger than the deadline).

On the other hand, with the other three approaches, schedulable solutions were produced, DP generating the smallest cost function followed in this order by MM and DM. The deviation of the MM approach from DP, calculated according to Equation 6.20, was of 2.44%, while for the DM approach the deviation was of 11.97% from DP.

Based on these results, and on the individual properties of each of the message scheduling approaches (see Section 6.4), the designer can decide which approach to use for the cruise controller implementation. However, the SM approach cannot be used for the vehicle cruise controller because it leads to an unschedulable system.

This chapter has constructed, step by step, a schedulability analysis for applications with data and control dependencies distributed on event-driven systems. The analysis will be used in the next chapter to determine the schedulability of the designs produced by a mapping and scheduling strategy that considers an incremental design process, in a fashion similar to our approach discussed in Chapter 5 for time-driven systems.

Chapter 7
Incremental Mapping for Event-Driven Systems

IN CHAPTER 5 we have discussed an incremental design strategy addressed to systems where *both* processes and messages are statically scheduled. However, as mentioned before, considering preemptive priority based scheduling for processes, with time triggered static scheduling for messages, can be the right solution under certain circumstances.

Hence, in this chapter we concentrate on scheduling and mapping of hard real-time systems which are implemented on distributed architectures. Process scheduling is based on a static priority preemptive approach while the bus communication is performed using the TTP. The mapping and scheduling tasks are considered in the context of an incremental design process as outlined in Section 2.2.

In Section 6.5 we have proposed four approaches for scheduling of messages using TTP that differ in the way the messages are allocated to the communication channel (either statically or dynamically) and whether they are split or not into packets for transmission. For each of these approaches, we have also developed a corresponding schedulability analysis.

Comparing these four approaches, in Section 6.8.2 we conclude that while the *Dynamic Packets Allocation* (DP) approach performs generally the best, since the dynamic scheduling of messages is able to reduce release jitter, but using the *Multiple Message Allocation* (MM) approach we can obtain almost the same result if the messages are carefully, off-line, allocated to slots. Moreover, in the case of larger process sets MM outperforms DP, as DP suffers from large overhead due to its dynamic nature. Also, DM performs worse than DP and MM because it does not split the messages into packets, and this results in a mismatch between the size of the messages dynamically queued and the slot size, leading to unused slot space that increases the jitter. SM performs the worst as it does not permit much room for improvement, leading to large amounts of unused slot space.

Therefore, for the purpose of this chapter, we consider that the messages are scheduled using the MM approach, and for the details of the corresponding schedulability analysis the reader is referred to Section 6.5. The discussion can easily be extended to any of the other three message passing approaches presented before.

The chapter is divided into five sections. The next two sections present some aspects of the general mapping and scheduling problem for event-driven systems, and the issues related to considering these design tasks within an incremental design process. Section 7.3 introduces the detailed problem formulation and the quality metrics we have defined. The mapping and scheduling strategy is presented in Section 7.4, and the approaches are evaluated in Section 7.5.

7.1 Application Mapping and Scheduling

In Section 5.1 of we have discussed some of the problems related to mapping and scheduling in the context of time-driven systems. In the beginning of this section we are going to have a look at the same design tasks, but in the context of event-driven systems, without considering, for the moment, an incremental design process. The particular issues related to mapping and scheduling for event-driven systems in the context of an incremental design approach will be presented later on in the discussion.

Example 7.1: Let us consider the example in Figure 7.1, where we have three system nodes N_1, N_2, N_3 (N_1 and N_3 having the same speed) and the TTP bus. Our task is to map P_1, P_2 and P_3 so that all deadlines are met. Process P_1 sends message m_1 to P_3 and message m_2 to P_2.

In the configuration presented in Figure 7.1a P_1 is mapped on N_3, P_3 on N_1 and P_2 is mapped on node N_2. Both m_1 and m_2 have to be sent in slot S_3 corresponding to node N_3 where the sender process P_1 is mapped. With the bus configuration such that m_1 is scheduled in the first round while m_2 is scheduled in the second round, P_2 misses its deadline (it has to wait for message m_2 sent by P_1).

However, with the same message schedule, if we map P_1 on N_1 and P_3 on N_3 as depicted in Figure 7.1b, m_1 and m_2 are sent in slot S_1 (corresponding to node N_1) which comes before S_3, and P_2 does not miss its deadline, receiving message m_2 on time. This could have also been achieved by a different scheduling of the messages presented in Figure 7.1c, where message m_2 is scheduled in the first round, and m_1 in the second, with the same mapping as in Figure 7.1a.

■

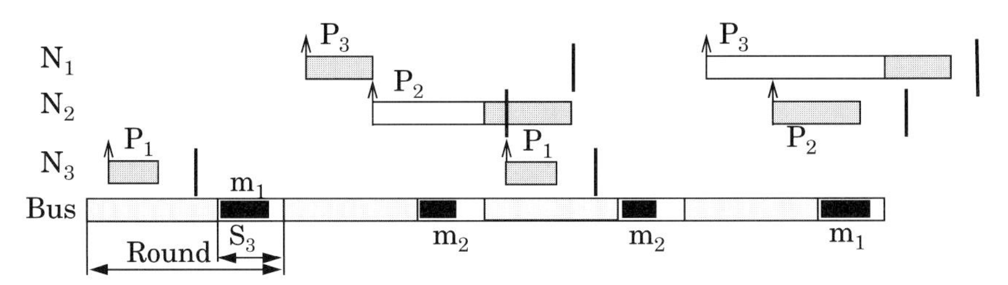

a) Processes P_1 on N_3, P_3 on N_1:
P_2 misses its deadline

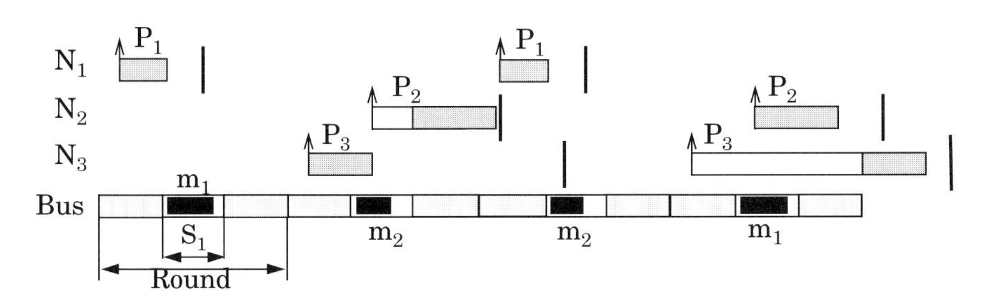

b) Processes P_1 on N_1, P_3 on N_3:
Processes P_2 mapped on N_2 meets its deadline

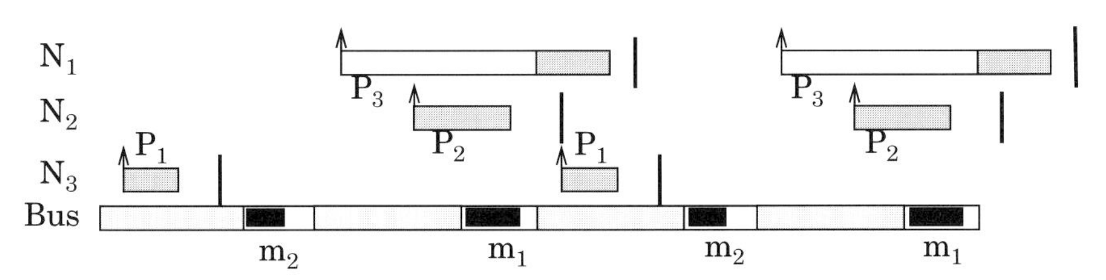

c) Same mapping as in case a,
message m_2 sent first: P_2 meets its deadline

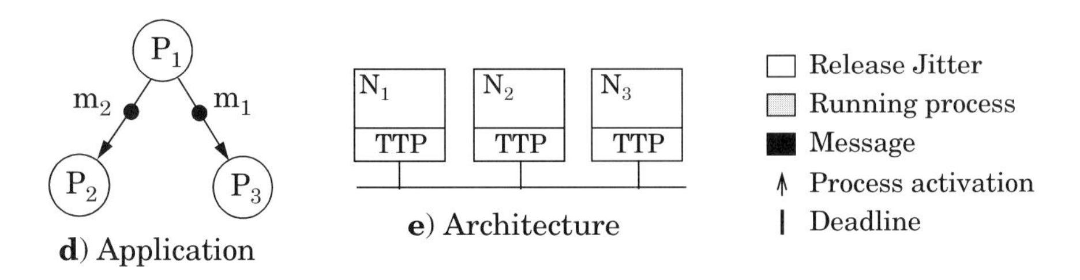

Figure 7.1: Mapping and Scheduling Example for Event-Driven Systems

However, as we have shown in Chapter 5 in the context of time-driven systems, it is not enough to produce a mapping and scheduling so that the system is schedulable if we are to support an incremental design process as discussed in Section 2.2.

Thus, we would like to perform mapping and scheduling such that: the timing constraints are satisfied, the modifications to the existing applications are minimized, and there is a good chance that future applications can easily be added to the resulted system.

Example 7.2: To illustrate the role of mapping and scheduling in the context of an incremental design process, let us consider the example in Figure 7.2, where we have two processors with the same speed connected by a TTP bus. With black we represent the set of already running applications ψ while the current application $\Gamma_{current}$ to be mapped and scheduled is represented in gray and consists of two processes and three messages.

In order for a system to be schedulable, a necessary condition is that the *utilization factor* of any node is less than one. We say that the processor can be "filled up" with processes until it reaches an utilization factor of one (the square depicting the processor is full). The utilization factor U_i of a process P_i is the ratio between the worst-case execution time C_i of that process and its period T_i: $U_i = C_i / T_i$. The utilization factor of a node is the sum of the utilization factors of all processes mapped on that node. The processes and messages that are to be mapped on the processors are depicted as blocks. The height of a process block is equal with its utilization factor, while the length of a message block gives the size of the message. White space on a processor represents available utilization, while white space on the bus represents available slack in the schedule table.

Now, let us suppose that, in the future, another application Γ_{future} has to be mapped on the system. Γ_{future} consists of two

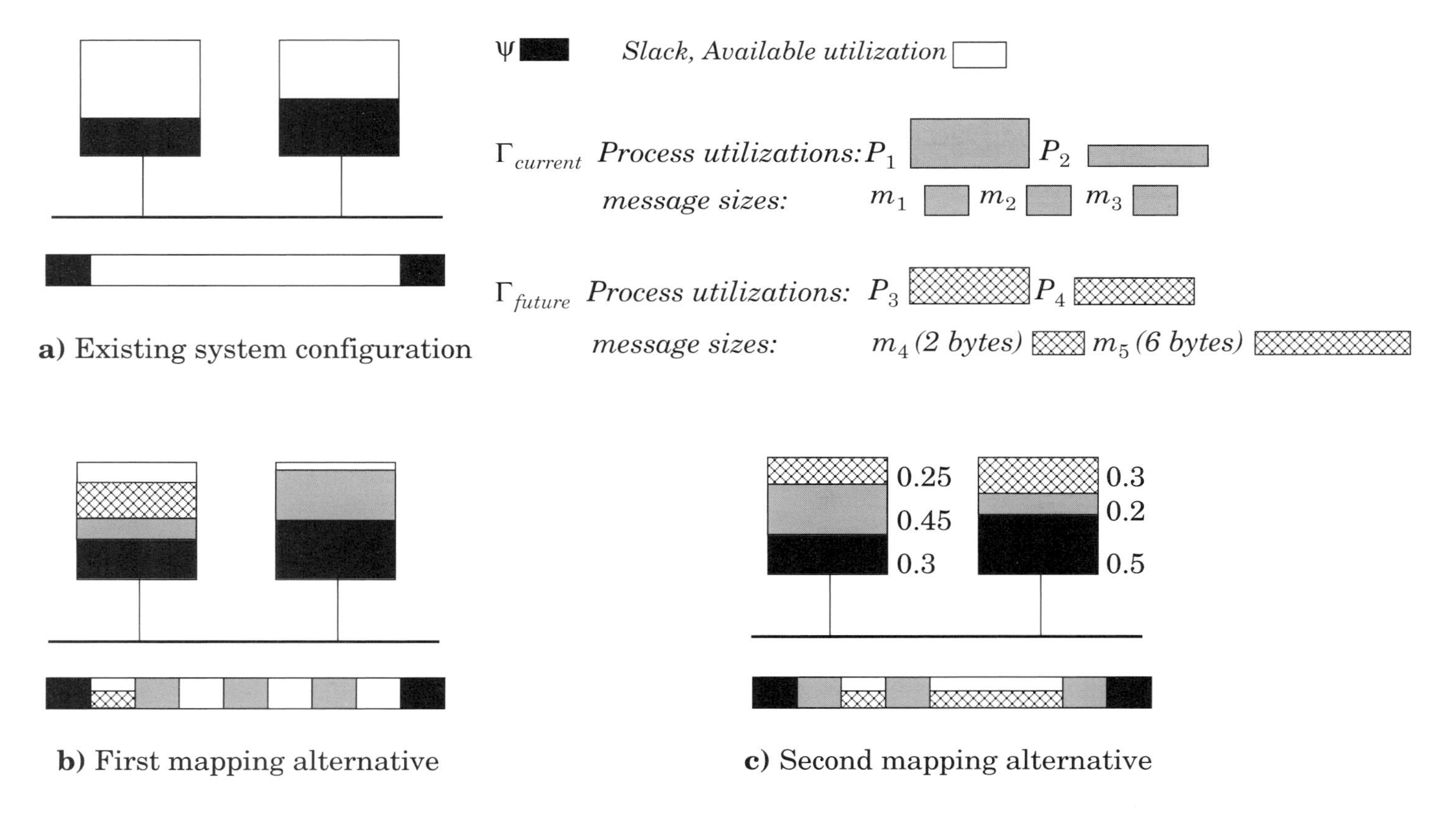

Figure 7.2: Incremental Mapping and Scheduling Example for Event-Driven Systems

processes and two messages represented as hashed blocks.

We can observe that the new application can be scheduled only in the second case, presented in Figure 7.2c. If $\Gamma_{current}$ has been implemented as in Figure 7.2b, we are not able to schedule process P_4 and message m_5 of Γ_{future}. The way our current application is mapped and scheduled will influence the likelihood of successfully mapping additional functionality on the system without being forced to modify the implementation of already running applications.

■

7.2 Mapping and Scheduling in an Incremental Design Approach

We model an application $\Gamma_{current}$ as a set of conditional process graphs, as outlined in Section 2.3.1. Thus, for each process P_i we know the set $\mathcal{N}_{P_i}$ of potential nodes on which it could be mapped and its worst-case execution time on each of these nodes. The underlying architecture is as presented in Section 3.4. We consider fixed priority preemptive scheduling for processes and a time-triggered message passing policy, as imposed by the TTP protocol.

Our goal is to map and schedule an application $\Gamma_{current}$ on a system that already implements a set ψ of applications, considering the following requirements, outlined previously in Chapter 5:

Requirement *a*	All constraints on $\Gamma_{current}$ are satisfied and minimal modifications are performed to the applications in ψ
Requirement *b*	New applications Γ_{future} can be mapped on top of the resulting system.

If it is not possible to map and schedule $\Gamma_{current}$ without modifying the already running applications, we have to change the mapping and scheduling of some applications in ψ. However, even with serious modifications performed on ψ, it is still possible that certain constraints are not satisfied. In this case the hardware architecture has to be changed by, for example, adding a new processor. Here we will not discuss this last case, but will concentrate on the situation where a possible mapping and scheduling which satisfies requirement *a* can be found, and this solution has to be further improved by considering requirement *b*.

In order to achieve our goals, we need certain information to be available concerning the set of applications ψ as well as the possible future applications Γ_{future}. In Section 2.3.2 we have presented the type of information we consider available for the existing applications in ψ, while in Section 5.2.1 we have shown how we can capture the characteristics of future time-driven applications. In the next section we outline the characterization of future event-driven applications. Moreover, as in the case of Chapter 5, we consider that $\Gamma_{current}$ can interact with the previously mapped applications ψ by reading messages generated on the bus by processes in ψ.

7.2.1 CHARACTERIZING FUTURE APPLICATIONS

In Section 5.2.1 we have argued that, given a certain limited application area (e.g., automotive electronics), it is possible to characterize the family of applications which in the future would be added to the current system.

Thus, we consider that, concerning the future applications, we know the set $S_U = \{U_{min}, ..., U_i, ..., U_{max}\}$ of possible *processor utilization factors* for processes, and the set $S_b = \{b_{min}, ..., b_i, ..., b_{max}\}$ of possible message sizes. The processor utilization factor U_i provides a measure of the computational load on a node N_i due to a process P_i, and is expressed as

$$U_i = \frac{C_i}{T_i}. \tag{7.1}$$

The utilization factors in S_U are considered relative to the slowest node in the system. All the other nodes are characterized by a speedup factor relative to this slowest node.

Thus, the utilization factor an entire application is given by

$$U = \sum_{i=1}^{n} U_i. \qquad (7.2)$$

We also assume that we know the distributions of probability $f_{S_U}(U)$ for $U \in S_U$ and $f_{S_b}(b)$ for $b \in S_b$.

Example 7.3: For example, we might have utilization factors $S_U = \{0.02, 0.05, 0.1, 0.2\}$ for the future application. If almost half of the processes are assumed to have an utilization factor of 0.1, and there is a lower probability of having processes with utilization factors of 0.2 and 0.02, then our distribution function $f_{S_U}(U)$ could look like this: $f_{S_U}(0.02) = 0.15$, $f_{S_U}(0.05) = 0.25$, $f_{S_U}(0.1) = 0.45$, $f_{S_U}(0.2) = 0.15$.

■

Another information is related to the period of future applications. In particular, the smallest expected period T_{min} is assumed to be given, together with the expected necessary bus bandwidth b_{need} inside such a period T_{min}. As will be shown later, this information is used in order to provide a fair distribution of slacks on the bus.

7.3 Quality Metrics and Exact Problem Formulation

Similarly to our approach to incremental mapping and scheduling for time-driven systems in Chapter 5, we develop two design criteria and associated metrics for event-driven systems, which are able to determine how well a system implementation supports an incremental design process.

We start by observing that a designer will be able to map and schedule an application Γ_{future} on top of a system implementing

ψ and $\Gamma_{current}$ only if there are sufficient resources available. In our case, the resources are the processor time and the bandwidth on the bus. In the context where processes are scheduled according to a fixed priority preemptive policy and messages are scheduled statically, having free resources translates into having enough processor capacity, and having space left for messages in the bus slots. We measure the processor capacity using the *available utilization*, while the available resources on the bus are called *slack*.

It is to be noted that the total quantity of computation and communication power available on our system after we have mapped and scheduled $\Gamma_{current}$ on top of ψ is the same regardless of the mapping and scheduling policies used. What depends on the mapping and scheduling strategy is the distribution of the available utilization on each processor, the size of the individual slacks on the bus, and the distribution of slacks along the time line. It is the distribution of available utilization and the size and distribution of the slacks that characterizes the quality of a certain design alternative. In this section we introduce the design criteria which reflect the degree to which one design alternative meets the requirement *b* introduced previously. For each criterion we provide metrics which quantify the degree to which the criterion is met. Relative to processes we have introduced one criterion which reflects how well the resulted available utilization on the nodes fits the requirements of a future application. For messages, there are two criteria. The first one reflects how well the resulted slack sizes fit a future application, and the second criterion expresses how well the slack is distributed over time.

7.3.1 PROCESSES RELATED CRITERION

The distribution of available utilization on the nodes, resulted after implementation of $\Gamma_{current}$ on top of ψ, should be such that it best accommodates a given family of applications Γ_{future}, characterized by the set S_U and the probability distribution f_{SU} as outlined before.

> **Example 7.4:** Let us consider the example in Figure 7.2, where we have two processors and the applications ψ and $\Gamma_{current}$ are already mapped. Suppose that application Γ_{future} consists of the two processes, P_3 and P_4. If we schedule $\Gamma_{current}$ like in Figure 7.2b it is impossible to fit Γ_{future} because there is not enough available utilization on any of the processors that can accommodate process P_4. A situation as the one depicted in Figure 7.2c is desirable, where the resulted available utilization is such that the future application can be accommodated.
>
> ■

In order to measure the degree to which the available utilization in a given design alternative fits the future applications, we provide a metric C_1^P which indicates to what extent the largest future application (considering the sum of available process utilization) could be mapped on top of the current design. This potentially largest application is determined knowing the total size of the available utilization, and the characteristics of the application: S_U and f_{SU}.

> **Example 7.5:** For example, if our *total* available utilization on *all* the processors is of 1.81 then we have to distribute this utilization according to the probabilities in f_{SU}. Considering the numerical example for processes given in Example 7.3, the largest application will be estimated to have a total of 20 processes: three processes of utilization 0.02, 5 of 0.05, 9 processes (almost half, $f_{SU}(0.1) = 0.45$) of utilization 0.1, and 3 of

0.2. If the number of processes for a particular dimension is not an integer, then we use the ceiling.

■

After we have determined the largest Γ_{future} we apply a *bin-packing* algorithm [Mar90] using the *best-fit* policy in which we consider processes as the objects to be packed, and the available utilization as containers. The total utilization of unpacked processes U_0 relative to the total utilization of the process set U_f gives the C_1^P metric: $C_1^P = (U_0 / U_f) \cdot 100$.

Example 7.6: In the case presented in Figure 7.2b $U_1 = 0.3$ and $U_2 = 0.25$, and P_2 represents 45% of the largest possible future application. In this case $C_1^P = 45\%$. However, in Figure 7.2c were we were able to completely map the future application $C_1^P = 0\%$.

■

7.3.2 CRITERIA RELATED TO MESSAGES

The first criterion for messages is similar to the one defined for processes. Thus, the slack sizes in the message schedule table MEDL (see Section 3.2.1) resulted after implementation of $\Gamma_{current}$ on top of ψ should be such that they best accommodate a given family of applications Γ_{future}, characterized by the set S_b and the probability distribution f_{S_b} for messages.

Example 7.7: Let us consider the example in Figure 7.2, where we have two processors and the applications ψ and $\Gamma_{current}$ are already mapped. Application Γ_{future} has two messages m_4 and m_5. It can be observed that the best configuration, taking into consideration only slack sizes, is to have a contiguous slack. However, in reality, it is almost impossible to map and schedule the current application such that a contiguous slack is obtained. Not only is it impossible, but it is also undesirable from the point of view of the second design criterion, discussed next. On the other side, as we can see from Figure 7.2b, if we schedule $\Gamma_{current}$ so that it fragments too much the slack, it is impossible to fit Γ_{future} because there is no slack that can accommodate message m_5. A situation as the one depicted in Figure 7.2c is desirable, where the resulted slack sizes can accommodate the characteristics of the Γ_{future} application.

■

In order to measure the degree to which the slack sizes in a given design alternative fit the future applications, we provide the metric C_1^m. The metric indicates how much of the communications of the largest future application which theoretically could be mapped on the system if the slacks on the bus would be summed, can be mapped on the current design alternative. The messages accounting for the largest amount of communication are determined, as shown above for processes, knowing the total size of the available slack, and the characteristics of the application: S_b and f_{S_b}.

C_1^m is calculated similarly to the metric C_1^P but, instead of packing the processes as objects, we try to pack the messages into the available slack on the bus. C_1^m is then the total size of unpacked messages, relative to the total size of messages in the largest future application.

Example 7.8: For Figure 7.2b, where m_5 could not be scheduled, C_1^m is 75% because m_4 of 6 bytes represents 75% of the total message sizes of 8 bytes. For the design alternative in Figure 7.2c C_1^m is 0% because all the messages have been scheduled.

■

We have just discussed a metric for how well the sizes of the slacks fit a possible future application. A similar metric is needed to characterize the distribution of slacks over time.

During implementation of $\Gamma_{current}$ we aim for a slack distribution such that the future application with the smallest expected period T_{min} and with the expected necessary bandwidth b_{need} inside the period T_{min}, can be accommodated. The minimum over the slacks inside each T_{min} period, which is available *periodically* to the messages of Γ_{future}, is the C_2^m metric.

Example 7.9: In Figure 7.3 we present a message schedule scenario. We consider a situation with T_{min} = 120 ms and b_{need} = 40 ms. The length of the schedule table is 360 ms, and the already scheduled messages of ψ and $\Gamma_{current}$ are depicted in black.

Let us consider the situation in Figure 7.3a. In the first period T_{min}, *Period 1*, there are 40 ms of slack available on the bus, in the second period 80 ms, and in the third period no slack is available. Hence, the total slack a future application with a period T_{min} can use on the bus in each period is C_2^m = min(40, 80, 0) = 0 ms. In this case, the messages cannot be scheduled. However, if we move m_1 to the left in the schedule table, we are able to create, in Figure 7.3b, 40 ms of slack in each period, resulting a C_2^m = 40 ms = b_{need}.

■

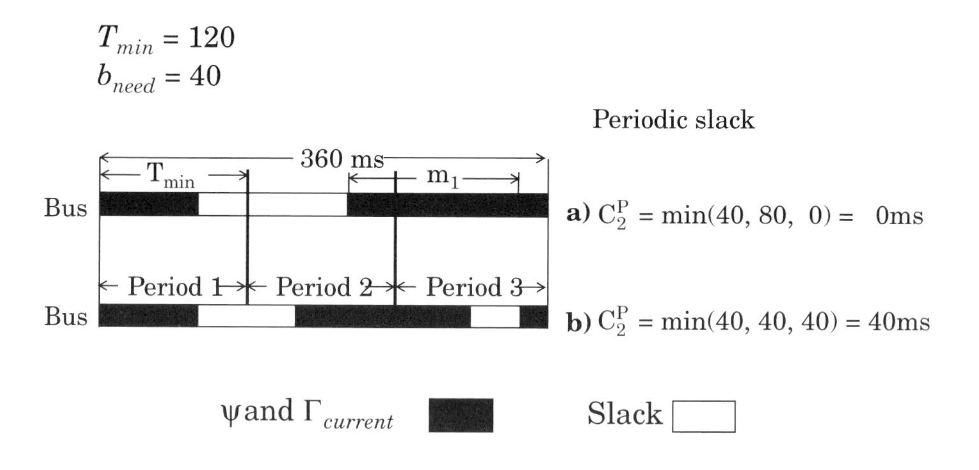

Figure 7.3: Example for the Second Message Design Criterion

7.3.3 COST FUNCTION AND EXACT PROBLEM FORMULATION

In order to capture how well a certain design alternative meets the requirement b stated previously, in Chapter 5 we have combined the design metrics in a cost function C. In the case of event-driven systems and the metrics presented in this chapter, the cost function is constructed similarly, as:

$$C = w_1^P (C_1^P)^2 + w_1^m (C_1^m)^2 + w_2^m max(0, b_{need} - C_2^m), \quad (7.3)$$

where the metric values are weighted by the constants w_i.

Our mapping and scheduling strategy will try to minimize this function. A design alternative that does not meet the second design criterion for messages is not considered a valid solution. Thus, using the last term, we strongly penalize the cost function if b_{need} is not satisfied, by using high values for the w_2^m weight.

At this point, we can give an exact formulation to our problem, which is synonymous with the problem addressed in Chapter 5 in the context of time-driven systems: Given an existing set of applications ψ which are already mapped and scheduled, and an application $\Gamma_{current}$ to be mapped on top of ψ, we are interested to find a mapping and scheduling of $\Gamma_{current}$ which satisfies all deadlines such that the existing applications are disturbed as little as possible. In our context, this means finding the subset $\Omega \subseteq \psi$ of old applications to be remapped and rescheduled such that we produce a valid solution for $\Gamma_{current} \cup \Omega$ and the total cost of modification $R(\Omega)$, as introduced in Section 2.3.2, is minimized. At the same time, the solution should minimize the cost function C, considering a family of future applications characterized by the sets S_U and S_b, the functions f_{SU} and f_{Sb} as well as the parameters T_{min} and b_{need}.

7.4 Mapping and Scheduling Strategy

The mapping and scheduling proposed in this section is similar to the one outlined in Section 5.4.2 for time-driven systems. The differences lie in the formulation of the design criteria and metrics, how are these used to select potential moves, and in the definition of the subset selection heuristics, which are tuned for event-driven systems.

As shown in Figure 7.4, our mapping and scheduling strategy (MH) has two main steps. In the first step we try to obtain a valid solution for $\Gamma_{current} \cup \Omega$ so that the total modification cost $R(\Omega)$ is minimized ($\Omega \subseteq \psi$ is the subset of existing applications that have to be modified to accommodate $\Gamma_{current}$). Starting from such a solution, a second step iteratively improves on the design in order to minimize the cost function C (Equation 7.3).

We iteratively improve the design using a transformational approach. A new design is obtained from the current one by performing a transformation called *move*. We consider the following moves: moving a process to a different node, and moving a message to a different slack on the bus. We only perform valid moves, which result in a schedulable system. The intelligence of the Mapping Heuristic lies in how the potential moves are selected. For each iteration a set of potential moves is generated by the PotentialMoveX functions. The SelectMoveX functions then evaluate these moves with regard to the respective metrics and selects the best one to be performed.

7.4.1 The Initial Mapping and Scheduling

The first step of MH consists of an iteration that tries subsets $\Omega \subseteq \psi$ with the intention to find that subset $\Omega = \Omega_{min}$ which produces a valid solution for $\Gamma_{current} \cup \Omega$ such that $R(\Omega)$ is minimized (lines 3–23 in Figure 7.4).

Given a subset Ω, the InitialMappingScheduling function (IMS) constructs a mapping and schedule for $\Gamma_{current} \cup \Omega$ that meets the

```
MappingSchedulingStrategy (MH)
1  Ω=∅
2
3  -- Step 1: try to find a valid schedule for Γcurrent that minimizes R(Ω)
4  repeat
5      succeeded=IMS(ψ\ Ω, Γcurrent ∪ Ω) -- initial mapping and scheduling
6      ASAP(Γcurrent ∪ Ω); ALAP(Γcurrent ∪ Ω)
7      -- compute worst case ASAP–ALAP intervals for messages
8      if succeeded then
9          -- try to satisfy the second message related design criterion
10         repeat
11             -- find moves with highest potential to maximize C2^m
12             move_set = PotentialMoveC2^m(Γcurrent ∪ Ω)
13             -- select and perform move which improves most C2^m
14             move = SelectMoveC2^m(move_set)
15             Perform(move)
16             succeeded = C2^m ≥ bneed
17         until succeeded or limit reached
18     end if
19     if succeeded and R(Ω) smallest so far then
20         Ωvalid = Ω; solutionvalid = solutioncurrent
21     end if
22     Ω = NextSubset(Ω) -- try another subset
23 until termination condition
24
25 if not succeeded then modify architecture; go to step 1; end if
26
27
28 -- Step 2: try to improve the cost function C
29 solutioncurrent = solutionvalid; Ωmin = Ωvalid
30 repeat -- find moves with highest potential to minimize C
31     move_set = PotentialMoveC1^P(Γcurrent ∪ Ωmin)
32         ∪ PotentialMoveC1^m(Γcurrent ∪ Ωmin)
33     -- select move which improves C and does not invalidate
34     -- the second message related design criterion
35     move = SelectMoveC1(move_set)
36     Perform(move)
37 until C1 has not changed or limit reached
38
end MappingSchedulingStrategy
```

Figure 7.4: The Mapping and Scheduling Strategy to Support Iterative Design

deadlines (both for processes in $\Gamma_{current}$ and those in Ω), without worrying about the design criteria in Section 7.3. For IMS we used as a starting point the mapping algorithm introduced in [Tin92], based on a simulated annealing strategy. We have modified the mapping algorithm in [Tin92] to consider during mapping a set of previous applications that have already been mapped, and to schedule the messages according to the TDMA protocol, using the MM approach (Section 6.5.2). The schedulability test that checks a particular mapping alternative is performed according to our schedulability analysis presented in Section 6.5.

If IMS succeeds in finding a mapping and a schedule which meet the deadlines, this is not yet a valid solution. In order to produce a valid solution we iteratively try to satisfy the second design criterion for messages (lines 10–17 in Figure 7.4). In terms of our metrics, that means a mapping and scheduling such that $C_2^m \geq b_{need}$. Potential moves can be the shifting of messages inside their worst case (largest) [*ASAP, ALAP*] interval in order to improve the periodic slack. In PotentialMoveC_2^m, line 12, we also consider movement of processes, trying to place the sender and receiver of a message on the same processor and, thus, reducing the bus load. SelectMoveC_2^m, line 14, evaluates these moves with regard to the second design criterion and selects the best one to be performed.

Example 7.10: Consider Figure 7.3a. In *Period 3* on node N_1 there is no available slack. However, if we move message m_1 with 40 ms to the left into *Period 2*, as depicted in Figure 7.3b, we create a slack in *Period 3*, thus the periodic slack on the bus will be min(40, 40, 40) = 40, instead of 0. ■

7.4.2 INCREMENTAL MAPPING AND SCHEDULING STRATEGY

If Step 1 of the MH algorithm (Figure 7.4) has succeeded, a mapping and scheduling of $\Gamma_{current} \cup \Omega$ has been produced which corresponds to a valid solution. In addition, Ω has the smallest minimization cost (minimization of the modification cost is introduced in Section 2.3.2 and detailed in Section 7.4.3). Starting from this valid solution, the second step of the MH strategy (lines 30–37) tries to improve on the design in order to minimize the cost function C. In a similar way as during Step 1, we iteratively improve the design by successive moves, without invalidating the second criterion achieved in the first loop.

The loop ends when there is no improvement to be achieved on the first two terms of the cost function, or a limit imposed on the number of iterations has been reached (line 37). For each iteration, the algorithm preforms moves that have the highest chance to improve the cost function. The moves are generated in the PotentialMoveC_1 functions (lines 31–32), and are evaluated and selected based on the respective metrics in the SelectMoveC_1 function (line 35). We now briefly discuss the PotentialMoveC_1^P and PotentialMoveC_1^m functions (PotentialMoveC_2^m has been discussed in the previous section).

PotentialMoveC_1^P

Let U_f be the total utilization factor of the largest future application Γ_{fmax}, and U_0 the utilization of that part which cannot be mapped in the current design alternative. This function is responsible for selecting moves of processes from one node to another so that $C_1^P = (U_0 / U_f) \cdot 100$ is reduced.

Moving a process P_i, with the utilization factor U_i, from a node N_j, where it is currently mapped, to a node N_k will increase the available utilization on node N_j to $U_{N_j} + U_i$, and decrease the available utilization on N_k to $U_{N_k} - U_i$. To find out U_0 in this new case would mean executing the bin-packing with the processes of the future application as objects and the new available utilization configuration as containers. This can take significant execution time since it has to be done for each potential move.

In Section 7.3 we have explained how we can estimate the processes that make up the largest future application Γ_{fmax} based on the total available utilization and the characterization of future applications. Let us assume that Γ_{fmax} consists of the set $\mathcal{P}_{fmax} = \{P_{f1}, P_{f2}, ..., P_{fn}\}$ of processes, and that $\mathcal{P}_0 = \{P_{fi}, P_{fi+1}, ..., P_{fm}\}$ are the ones that cannot be mapped in the current design alternative. The total utilization requested by the unmapped processes is $U_0 = U_{fi} + U_{fi+1} + ... + U_{fm}$. For the potential move of P_i from N_j to N_k we have to recalculate C_1^P which means determining U_0.

In order to reduce the execution time needed by the bin-packing algorithm, we do not consider all the processes of Γ_{fmax} as objects to be packed. We consider for repacking only those processes belonging to Γ_{fmax} that had to be removed from N_k to make room for P_i, together with those that were already left outside. Our heuristic considers that to make room for P_i on node N_k we remove those processes $\mathcal{P}_i^{N_k} \subset \Gamma_{fmax}$ mapped on N_k which have the smallest utilization factor, since they are the ones that should be easiest to fit on other nodes. The metric used by SelectMoveC$_1$ to rank this move is the sum of the utilization factors of processes which are left out after trying to repack the $\mathcal{P}_0 \cup \mathcal{P}_i^{N_k}$ set.

Out of the best moves according the previous metric, we encourage those that have the smallest impact on the schedulability analysis, since we would like to keep the system schedulable. This means moving processes that have low priority (do not have a large impact on other processes) and have a response time that is considerably smaller than their deadline ($D_i - R_i$ is large).

PotentialMoveC_1^m

In order to avoid excessive fragmentation of the slack on the bus we will consider moving a message to a position that "snaps" to another existing message. A message is selected for potential move if it has the smallest "snapping distance," i.e., in order to attach it to other message it has to travel the smallest distance inside the schedule table. We also consider moves that try to increase the individual slacks sizes. Therefore, we first eliminate slack that is unusable: it is too small to hold the smallest message of the future application. Then, the slacks are sorted in ascending order and the smallest one is considered for improvement. Such improvement of a slack is performed through moving a nearby message, but avoiding to create as a result an even smaller individual slack.

7.4.3 MINIMIZING THE MODIFICATION COST

In the first step of our mapping strategy, described in Figure 7.4, we iterate on subsets Ω to search for a valid solution which also minimizes the total modification cost $R(\Omega)$. As a first attempt, the algorithm searches for a valid implementation of $\Gamma_{current}$ without disturbing the existing applications ($\Omega = \varnothing$). If no valid solution is found successive subsets Ω produced by the function NextSubset are considered, until a terminating condition is met.

In Section 5.4.3 of Chapter 5 we have presented several approaches to the implementation of the NextSubset function in the context of time-driven systems. The same strategies are used in this chapter, but now in the case of event-driven systems. The difference lies in the formulation of the Δ metrics which guide the subset selection process.

The first approach to the implementation of the NextSubset function is an exhaustive search algorithm (ES), similar to the one presented in Section 5.4.3. As shown in that chapter, the exhaustive approach that finds an optimal solution can be used only for small sets of applications. The second approach presented in Section 5.4.3 is a greedy heuristic, here named *Ad-hoc Subset Selection* (AS), which finds very quickly a valid solution, if one exists, with the drawback that the corresponding total modification cost is higher than the optimal one. However, as we argue in Chapter 5 an intelligent heuristic should be able to identify the reasons due to which a valid solution has not been found and use this information when selecting applications to be included in Ω The next section presents such a heuristic in the case of event-driven systems.

Subset Selection Heuristic (SH)

There can be two possible causes for not finding a valid solution: an initial mapping which meets the deadlines has not been produced, or the second criterion is not satisfied.

Let us investigate the first reason. If an application Γ_i is schedulable, this means that all its processes meet their deadlines. If IMS determines that the application is not schedulable this means that at least one of the processes P_i missed its deadline: $R_i > D_i$. Besides the intrinsic properties of the application that can lead to this situation, process P_i can miss its deadline also because of the interference of higher priority processes that are mapped on the same node with P_i, processes that can also belong to other applications. In this situation we say that there is a *conflict* with processes belonging to other applications. We are interested to find out which applications are responsible for conflicts encountered by our $\Gamma_{current}$, and not only that, but also which ones are *flexible* enough to move away in order to avoid these conflicts ($D_i - R_i$ is large).

IMS determines a metric Δ_i that characterizes the degree of conflict and the flexibility of application Γ_i in relation to $\Gamma_{current}$. A set of applications Ω will be characterized, in relation to $\Gamma_{current}$, by:

$$\Delta(\Omega) = \sum_{\Gamma_i \in \Omega} \Delta_i . \tag{7.4}$$

The metric $\Delta(\Omega)$ will be used by our subset selection heuristic if IMS has failed to produce a solution which satisfies the deadlines. An application with a larger Δ_i is more likely to lead to a valid schedule if included in Ω.

Basically, Δ_i is the total amount of *interference* caused by higher priority processes of Γ_i to processes in $\Gamma_{current}$. For a process P_i, the interference I_{ji} from a higher priority process P_j mapped on the same node, is the time that P_j delays the execution of P_i, and is given by:

$$I_{ji} = \left\lceil \frac{J_j + R_i}{T_j} \right\rceil C_j \tag{7.5}$$

where J_j is the release jitter of process P_j and a detailed description of how it is calculated in the context of the MM approach for message scheduling over TTP is given in Section 6.5.2. Figure 7.5 presents in more detail how Δ_i is calculated.

If the initial mapping was successful, the first step of MH could fail during the attempt to satisfy the second design criterion for messages. In this case, the metric Δ_i is computed in a different way. It will capture the potential of an application Γ_i to improve the metric C_2^m if remapped together with $\Gamma_{current}$. Thus, for the improvement of C_2^m we consider a total number of moves from all the non-frozen applications (determined using PotentialMove$C_2^m(\psi)$, see Section 7.4.2). For each move that has as subject $m_j \in \Gamma_i$, we increment the metric Δ_i with the predicted improvement on C_2^m.

MH starts by trying an implementation of $\Gamma_{current}$ with $\Omega = \varnothing$. If this attempt fails, because of one of the two reasons mentioned above, the corresponding metrics Δ_i are computed for all $\Gamma_i \in \psi$. Our heuristic SH will then start by finding the ad-hoc solution Ω_{AS} produced by the AS algorithm (this will succeed if there

```
DeltaMetrics(Γcurrent, Ω)
1   for each non frozen Γi ∈ Ω do
2       Δi = 0
3   end for
4
5   for each Pi ∈ Γcurrent do
6       if Ri > Di then
7           for each non frozen Γk ∈ Ω do
8               -- hp(Pi) is the set of processes with higher priority than Pi
9               for each Pj ∈ Γk ∩ hp(Pi) do
10                  Δk = Δk + Cj * ⌈(Jj + Ri) / Tj⌉
11              end for
12          end for
13      end if
14  end for
15
16  return Δ
end DeltaMetrics
```

Figure 7.5: Determining the Delta Metrics

exists any solution) with a corresponding cost $R_{AS} = R(\Omega_{AS})$ and a $\Delta_{AS} = \Delta(\Omega_{AS})$. SH now continues by trying to find a solution with a more favorable Ω (a smaller total cost R). Therefore, the thresholds $R_{max} = R_{AS}$ and $\Delta_{min} = \Delta_{AS}/n$ (for our experiments we considered $n = 2$) are set. For generating new subsets Ω, the function NextSubset now follows a similar approach like ES but in a reverse direction, towards smaller subsets, and it will consider only subsets with a smaller total cost than R_{max} and a larger Δ than Δ_{min} (a small Δ means a reduced potential to eliminate the cause of the initial failure). Each time a valid solution is found, the current values of R_{max} and Δ_{min} are updated in order to further restrict the search space. The heuristic stops when no subset can be found with $\Delta > \Delta_{min}$, or a certain imposed limit has been reached (e.g., on the total number of attempts to find new sets).

7.5 Experimental Evaluation

For the evaluation of our mapping strategies we first used applications containing 40, 80, 160, 240 and 320 processes representing the $\Gamma_{current}$ application generated for experimental purpose. Thirty applications were generated for each dimension, thus a total of 150 applications were used for experimental evaluation. We considered an architecture consisting of 10 nodes of different speeds. For the communication channel we considered a transmission speed of 256 Kbps and a length below 20 meters. The maximum length of the data field in a bus slot was 8 bytes. All experiments were run on a SUN Ultra 10.

7.5.1 MODIFICATION COST MINIMIZATION HEURISTICS

The first result concerns the quality of the designs obtained with our mapping strategy MH using the search heuristic SH compared to the case when the ad-hoc approach AS and the exhaustive search ES are used for subset selection.

For each of the five application dimensions generated we have considered a set of existing applications ψ consisting of 160, 240, 320, 400 and 480 processes, respectively. The sets contained 4, 6, 8, 10 and 12 applications, each application with an associated modification cost assigned manually in the range 10 to 100. The dependencies between applications were such that the total number of subsets resulted for each set ψ were 8, 32, 128, 256, and 1024. We have considered that the future applications Γ_{future} consist of a process set of 80 processes, randomly generated according to the following specifications: S_U = {0.02, 0.05, 0.1, 0.15, 0.2}, $f_{S_U}(S_U)$ = {0.1, 0.25, 0.45, 0.15, 0.05}, S_b = {2, 4, 6, 8 bytes}, $f_{S_b}(S_b)$ = {0.2, 0.5, 0.2, 0.1}, T_{min} = 250 ms, and b_{need} = 20 ms.

MH has been used to produce a valid solution for each of the 150 process sets representing $\Gamma_{current}$ on top of the existing applications ψ using the ES, AS and SH approaches to subset selection.

For each of the resulted valid solutions, there corresponds a minimum modification cost $R(\Omega_{min})$. Figure 7.6a compares the three approaches to subset selection based on the modification cost needed in order to obtain a valid solution. The exhaustive approach ES is able to obtain valid solutions at the optimum (smallest) modification cost, (e.g., less than 400, in average, for systems with 12 applications consisting of a total of 480 processes), while the ad-hoc approach AS needs in average 3.11 times more costly modifications in order to obtain valid solutions (e.g., more than 1100 for 480 processes in Figure 7.6a). However, in order to find the optimal re-mapping the ES approach needs large computation times. For example, it can take more than 35 minutes, in average, in order to find the smallest cost subset to be remapped that leads to a valid solution in the case we have 12 applications (corresponding to 480 processes in Figure 7.6b). From Figure 7.6 we can see that the proposed heuristic SH performs quite well, needing only 1.84 times larger costs, in average, in order to obtain a valid schedule at a computation cost comparable with the fast ad-hoc approach AS (see Figure 7.6b). For the results in Figure 7.6 we have eliminated those situations in which a valid solution has not been produced by MH (which means that there is no solution regardless of the modification cost).

7.5.2 INCREMENTAL MAPPING AND SCHEDULING HEURISTICS

Next, we were interested to investigate the quality of the mapping heuristic MH compared to a so called *ad-hoc mapping approach* (AM).

To concentrate on this, we have considered that *no modifications* are allowed to the applications in ψ. The AM approach is a simple, straightforward solution to produce designs which, to a certain degree, support an incremental process. AM tries to evenly balance the available utilization remaining after mapping the current application. The quality of the designs obtained

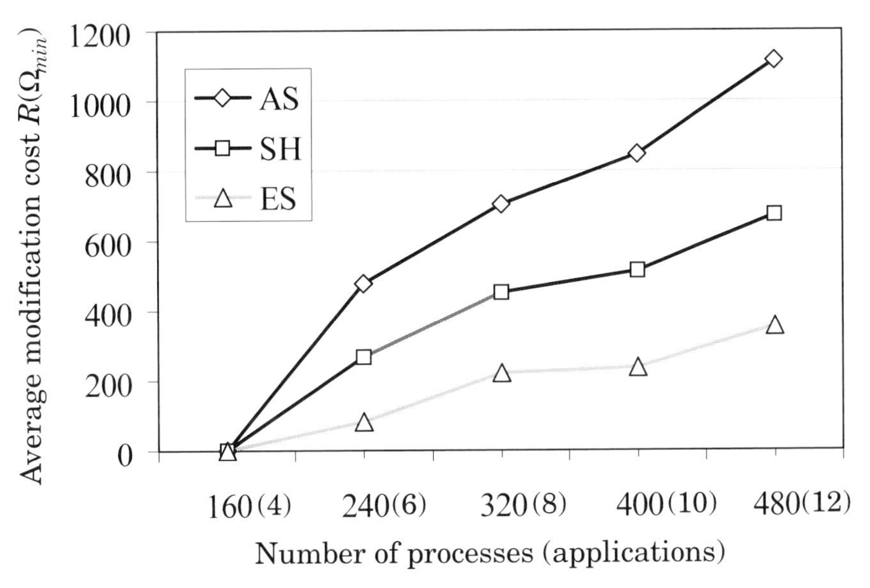

a) Average modification costs for AS, SH, ES

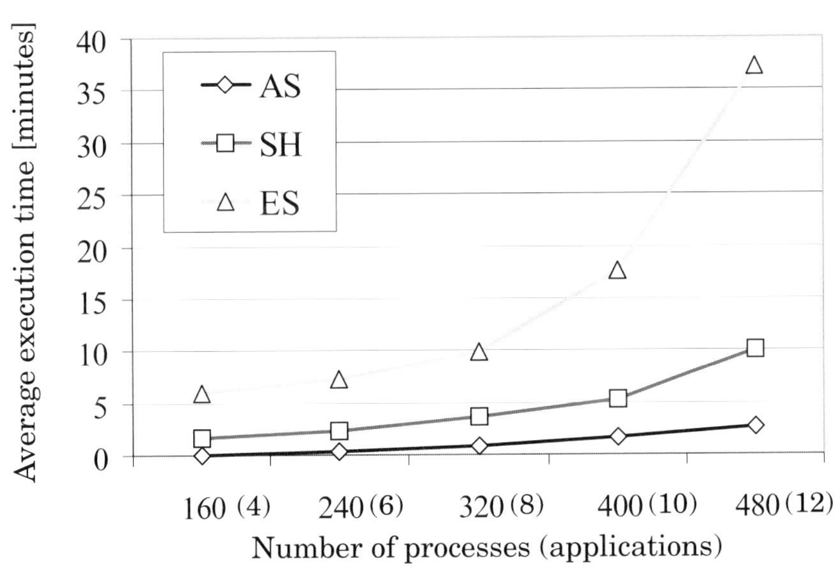

b) Execution times for AS, SH, ES

Figure 7.6: Evaluation of the Modification Cost Minimization Heuristics

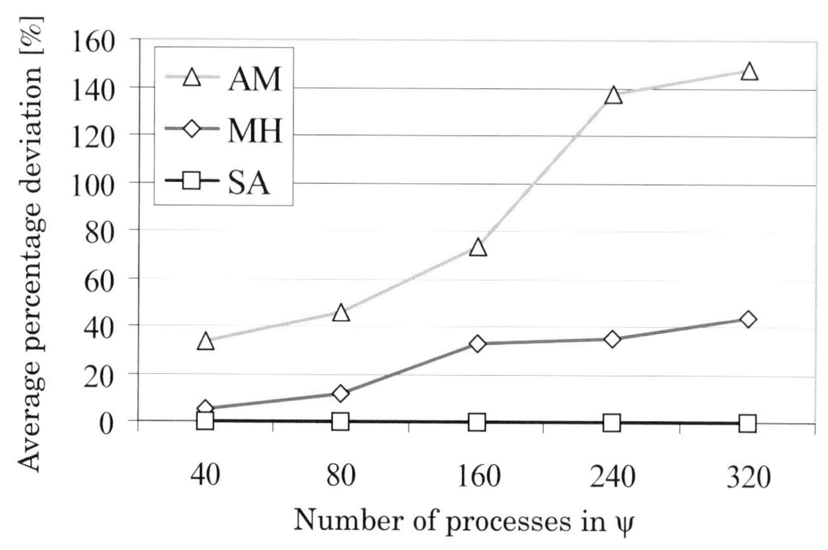

a) Percentage deviations for AM, MH, SA

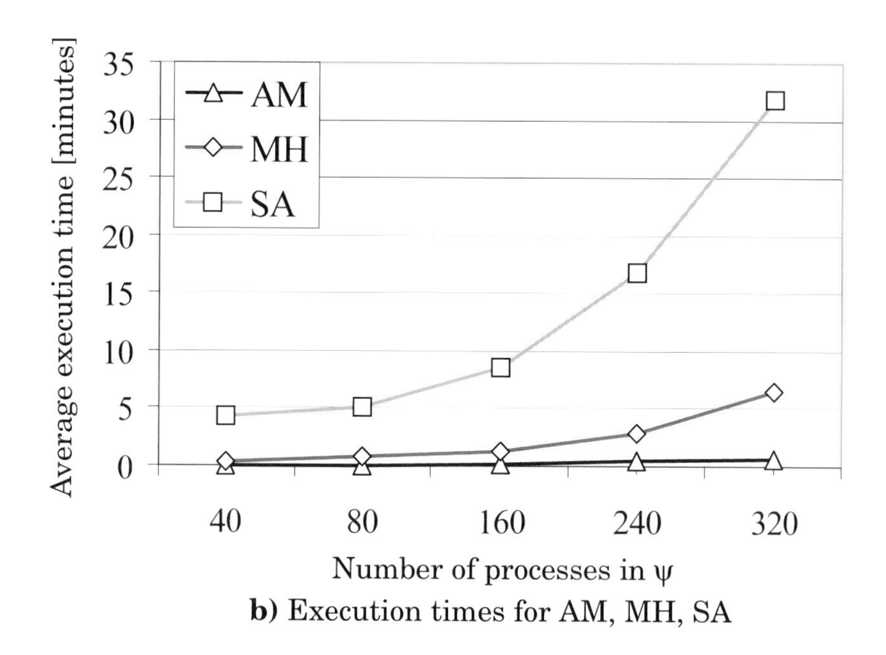

b) Execution times for AM, MH, SA

Figure 7.7: Evaluation of the Design Transformation Heuristics

with MH and AM were compared with a near-optimal mapping and schedule obtained with a Simulated Annealing strategy (SA) strategy (Appendix A), that minimizes the cost function C (Section 7.3.3). One of the drawbacks of the SA strategy is that in order to find near-optimal solutions it needs very large computation times. Such a strategy, although useful for the final stages of the system synthesis, cannot be used inside a design space exploration cycle.

MH, SA and AM have been used to map each of the 150 applications representing $\Gamma_{current}$ on the existing applications ψ For each of the resulted designs, the objective function C has been computed. Very long and expensive runs have been performed with the SA algorithm for each process set and the best ever solution produced has been considered as the near-optimum for that process set. We have compared the cost function obtained for the 150 applications considering each of the three mapping algorithms. Figure 7.7a presents the average percentage deviation of the cost function obtained with the MH and AM from the value of the cost function obtained with the near-optimal scheme. We have excluded from the results in Figure 7.7, 28 solutions obtained with AM for which the second design criterion for messages has not been met, and thus the objective function has been strongly penalized. The average run-times of the algorithms, in minutes, are presented in Figure 7.7b. The SA approach performs best in terms of quality at the expense of a large execution time. The execution time can be up to 40 minutes for large applications of 320 processes. MH performs very well, and is able to obtain good quality solutions in a very short time. AM is very fast, but since it does not address explicitly the design criteria presented in Section 7.2 it has the worst quality of solutions, according to the cost function.

The most important aspect of the experiments is determining to which extent the mapping strategies proposed in this chapter really facilitate the implementation of future applications. To find this out, we have mapped applications of 40, 80, 160 and

240 processes representing the $\Gamma_{current}$ application on top of the previously generated existing applications ψ After mapping and scheduling each of these applications we have tried to add a new application Γ_{future} to the resulted system. Γ_{future} consists of 80 processes, randomly generated according to the same specifications presented before. The experiments have been performed two times, using first MH*[1] and then AM for mapping $\Gamma_{current}$. In both cases we were interested if it is possible to find a valid implementation for Γ_{future} on top of $\Gamma_{current}$ using the initial mapping algorithm IMS. Figure 7.8a shows the number of successful implementations in the two cases. In the case $\Gamma_{current}$ has been mapped with MH*, this means using the design criteria and metrics proposed in this chapter, we were able to find a valid schedule for 56% of the total mapping attempts with IMS using Γ_{future}. However, using AM to map $\Gamma_{current}$ has led to a situation where IMS is able to find valid schedules in only 31% of the cases.

Another observation from Figure 7.8 is that when the available utilization is large, as in the case $\Gamma_{current}$ has only 40 processes, it is easy for both MH* and AM to find a mapping that allows adding future applications. However, as $\Gamma_{current}$ grows to 80, only MH* is able to find a mapping of $\Gamma_{current}$ that supports an incremental design process, accommodating more than 60% of the future applications, while using AM only less than 25% are accommodated. If the remaining utilization is very small, after we map a $\Gamma_{current}$ of 240, it becomes practically impossible to map new applications without modifying the current system.

However, in the case the mapping heuristic is *allowed to modify* the existing system as discussed in this chapter then we are able to increase the number of successfully mapped Γ_{future} applications to 73% from the total instead of only 56%. The percentage of accommodated Γ_{future} applications, for different dimensions of $\Gamma_{current}$, if modifications are allowed on the exist-

1. MH* is the same mapping heuristic as in Figure 7.4, but in which we do not allow modifications to the existing applications.

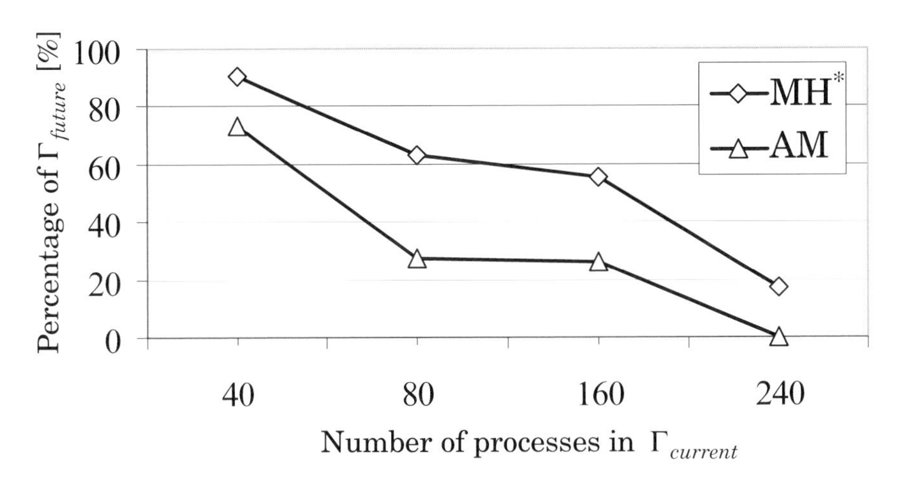

a) Percentage of Γ_{future} applications successfully mapped, no modifications allowed

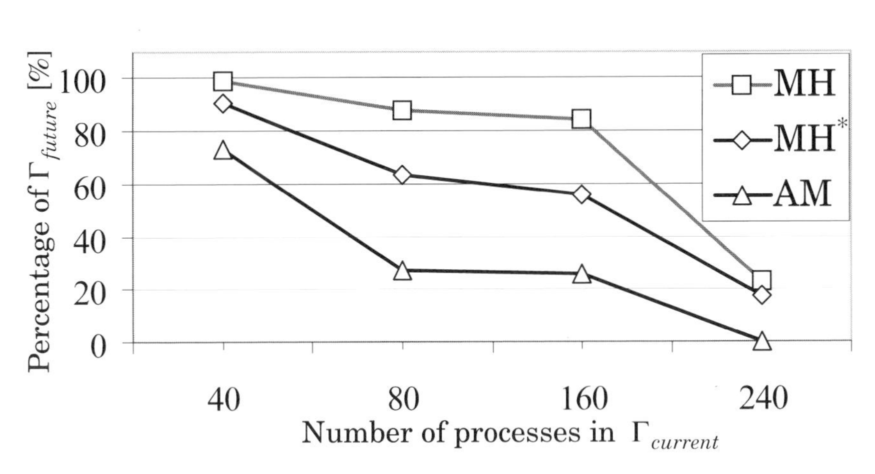

b) Percentage of Γ_{future} applications successfully mapped, modifications allowed

Figure 7.8: Percentage of Future Applications Successfully Mapped

ing system, is shown by the diagram MH in Figure 7.8b. After mapping a $\Gamma_{current}$ with 80 processes using MH we are able to accommodate 88% of the future applications, compared to only 61% in the case we do not allow modifications to the existing system (MH^*). Such an increase is, of course, expected. The important aspect, however, is that it is obtained not by randomly selecting old applications to be remapped, but by performing this selection such that the total modification cost is minimized.

7.5.3 THE VEHICLE CRUISE CONTROLLER

Finally, we considered an example implementing a vehicle cruise controller (CC):

- The CC has 32 processes, and is modeled as an un-mapped conditional process graph, presented in Figure 2.9 on page 40.
- The cruise controller is to be mapped on an architecture consisting of 5 nodes, interconnected by TTP, as presented in Figure 2.7a on page 37.
- The software architecture for event-triggered systems used by the CC is introduced in Section 3.4.

The system ψ consists of 80 processes generated randomly. The CC is the $\Gamma_{current}$ application to be mapped. We have also generated 30 future applications of 40 processes each with the characteristics of the CC, which are typical for automotive applications. By mapping the CC using MH^* we were able to later map 18 of the future applications, while using AM only 6 of the future applications could be mapped. MH^* and AM do not allow modifications of the existing system. When modifications are allowed, using the MH approach, we are able to map 26 of the 30 future applications.

As the experiments have shown, the design criteria proposed in this chapter, for event-driven systems, are able to drive the optimization process towards solutions that support an incremental design process.

This and the previous part of the book have addressed ET and TT systems, respectively. In the next part, we will consider multi-cluster systems, designed as interconnected clusters of processors, where each such cluster can be either TT or ET.

PART IV
Multi-Cluster Systems

Chapter 8
Schedulability Analysis and Bus Access Optimization for Multi-Cluster Systems

THIS CHAPTER PRESENTS an approach to schedulability analysis and bus access optimization for multi-cluster distributed embedded systems consisting of time-triggered and event-triggered clusters, interconnected via gateways, as introduced in Section 3.5.

On the time-triggered clusters (TTC) the processes are scheduled based on a non-preemptive static cyclic scheduling policy, and messages are sent using the TTP, while on the event-triggered clusters (ETC) we use a fixed-priority preemptive scheduling policy for processes, and messages are sent via the CAN bus.

We have proposed a schedulability analysis for multi-cluster systems, including a buffer size and worst case queuing delay analysis for the gateways, responsible for routing inter-cluster traffic. Optimization heuristics for the priority assignment and synthesis of bus access parameters aimed at producing a schedulable system with minimal buffer needs have also been developed.

This chapter is organized in five sections. The next section introduces the problems that we are addressing in this chapter. Section 8.2 presents our proposed schedulability analysis for multi-cluster systems, and Section 8.3 uses this analysis to drive the optimization heuristics used for system synthesis. The last section present the experimental results.

8.1 Problem Formulation

As input to our problem we have an application Γ given as a set of conditional process graphs mapped on an architecture consisting of a TTC and an ETC interconnected through a gateway node. The set of nodes on the TTC is denoted with $\mathcal{N}_T$, the ETC consists of the set of nodes $\mathcal{N}_E$, and the gateway node is denoted with N_G.

We are interested first to find a system configuration denoted by a 3-tuple $\psi = <\phi, \beta, \pi>$ such that the application Γ is schedulable. Determining a system configuration ψ means deciding on:

- The set ϕ of the offsets corresponding to each process and message in the system (see Section 6.2). The offsets of processes and messages on the TTC practically represent the local schedule tables and MEDLs.
- The TTC bus configuration β, indicating the sequence and size of the slots in a TDMA round on the TTC.
- The priorities of the processes and messages on the ETC, captured by π.

Once a configuration leading to a schedulable application is found, we are interested to find a system configuration that minimizes the total queue sizes needed to run a schedulable application. The approach presented in this chapter can be extended to cluster configurations where there are several ETCs and TTCs interconnected by gateways.

Example 8.1: Let us consider the example in Figure 8.1 where we the application G_1 mapped on the a two-cluster

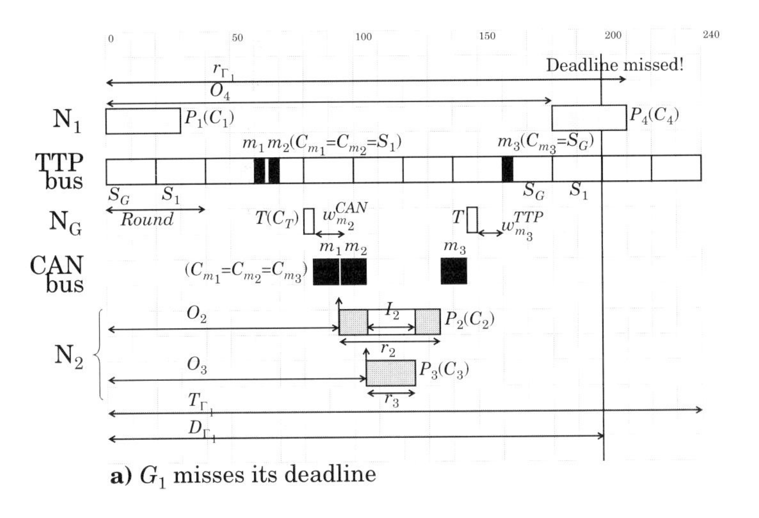

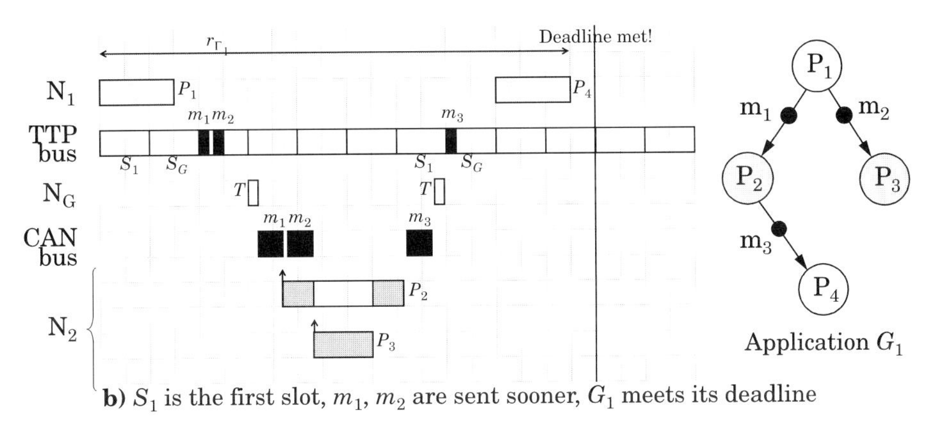

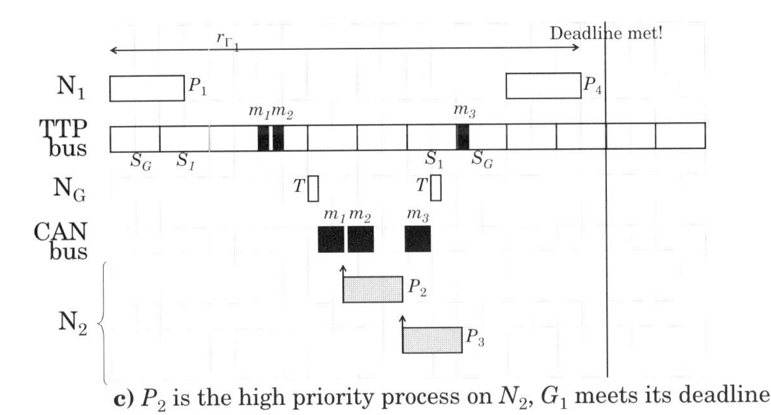

Figure 8.1: Scheduling Examples for Multi-Clusters

system as illustrated in Figure 3.8 on page 60. In the system configuration of Figure 8.1 we consider that, on the TTP bus, the gateway transmits in the first slot (S_G) of the TDMA round, while node N_1 transmits in the second slot (S_1). The priorities inside the ETC have been set such that $priority_{m_1} > priority_{m_2}$ and $priority_{P_3} > priority_{P_2}$.

In such a setting, G_1 will miss its deadline, which was set at 200 ms. However, changing the system configuration as in Figure 8.1b, so that slot S_1 of N_1 comes first, we are able to send m_1 and m_2 sooner, and thus reduce the response time and meet the deadline. The response times and resource usage do not, of course, depend only on the TDMA configuration. In Figure 8.1c, for example, we have modified the priorities of P_2 and P_3 so that P_2 is the higher priority process. In such a situation, P_2 is not interrupted when the delivery of message m_2 was supposed to activate P_3 and, thus, eliminating the interference, we are able to meet the deadline, even with the TTP bus configuration of Figure 8.1a.

■

8.2 Multi-Cluster Scheduling

In this section we propose an analysis for hard real-time applications mapped on multi-cluster systems. The aim of such an analysis is to find out if a system is schedulable, i.e., all the timing constraints are met. In addition to this, we are also interested to bound the queue sizes needed to run a schedulable applications.

On the TTC, an application is schedulable if it is possible to build a schedule table such that the timing requirements are satisfied. On the ETC, the answer whether or not a system is schedulable is given by a schedulability analysis, and we use the schedulability analysis outlined in Section 6.4.1.

In Section 6.4.1 the release jitter of a destination process D depends on the communication delay between sending and receiving an incoming message m: $J_{D(m)} = r_m$. However, in the case of a multi-cluster system, we will use offsets to capture the communication delays, and not the release jitter. Thus, the offset of a process will be determined such that it contains the communication delay due to the incoming message. For example, the offset O_2 of process P_2 in Figure 8.1a has been set such that it accounts for the delay due to message m_1 sent from the gateway transfer process T to the process P_2 via the CAN bus.

Moreover, determining the schedulability of an application mapped on a multi-cluster system cannot be addressed separately for each type of cluster, since the inter-cluster communication creates a circular dependency: the static schedules determined for the TTC influence through the offsets the response times of the processes on the ETC, which on their turn influence the schedule table construction on the TTC.

> **Example 8.2:** In Figure 8.1a, placing m_1 and m_2 in the same slot leads to equal offsets for P_2 and P_3. Because of this, P_3 will interfere with P_2 (which would not be the case if m_2 sent to P_3 would be scheduled in *Round 4*) and thus the placement of P_4 in the schedule table has to be accordingly delayed to guarantee the arrival of m_3.
>
> ■

In our response time analysis we consider the influence between the two clusters by making the following observations:

- The start time of process P_i in a schedule table on the TTC is its offset O_i.
- The worst-case response time r_i of a TT process is its worst-case execution time, i.e. $r_i = C_i$ (TT processes are not preemptable).
- The worst-case response times of the messages exchanged between two clusters have to be calculated according to the schedulability analysis to be described in Section 8.2.1.

- The offsets have to be set by a scheduling algorithm such that the precedence relationships are preserved. This means that, if process P_j depends on process P_i, the following condition must hold: $O_j \geq O_i + r_i$. Note that for the processes on a TTC receiving messages from the ETC this translates to setting the start times of the processes such that a process is not activated before the worst-case arrival time of the message from the ETC. In general, offsets on the TTC are set such that all the necessary messages are present at the process invocation.

The MultiClusterScheduling algorithm in Figure 8.2 receives as input the application Γ, the mapping M, the system configuration ψ and produces the offsets ϕ and worst-case response times ρ.

The algorithm sets initially all the offsets to 0 (line 2). Then, the worst-case response times are calculated using the ResponseTimeAnalysis function (line 5) using the feasible analysis provided in [Tin94b]. The fixed-point iterations that calculate the response times at line 4 will converge if processor and bus loads are smaller than 100% [Tin94b]. Based on these worst-case response times, we determine new values ϕ^{new} for the offsets using a list scheduling algorithm (line 7).

The multi-cluster scheduling algorithm loops until the degree of schedulability δ_Γ of the application Γ cannot be further reduced (lines 9–22). In each loop iteration, we select a new offset O_i from the set of ϕ^{new} offsets (line 10), and run the response time analysis (line 12) to see if the degree of schedulability has improved (line 13). That offset O_i is selected, which corresponds to the unschedulable process P_i (i.e., its worst-case response time r_i is greater than its deadline D_i) with the largest difference $r_i - D_i$. If δ_Γ has not improved, we continue with the next offset in ϕ^{new}.

When a new offset O_i^{new} leads to an improved δ_Γ, we exit the for-each loop 10–21 that examines offsets from ϕ^{new}. The loop

```
MultiClusterScheduling(Γ, M, ψ)
1  -- determines the set of offsets φ and worst-case response times ρ
2  for each O_i ∈ φ do O_i = 0 end for -- initially all offsets are zero
3  -- determine initial values for the worst-case response times
4  -- according to the analysis in Section 8.2.1
5  ρ = ResponseTimeAnalysis(Γ, M, ψ φ)
6  -- determine new values for the offsets, based on ρ
7  φ^new = ListScheduling(Γ, M, ψ ρ)
8  δ_Γ = ∞-- consider the system unschedulable initially
9  repeat -- iteratively improve the degree of schedulability δ_Γ
10     for each O_i^new ∈ φ^new do -- for each newly calculated offset
11         O_i^old = φ.O_i; φ.O_i = φ^new.O_i^new -- set the new offset, remember old
12         ρ^new = ResponseTimeAnalysis(Γ, M, ψ φ)
13         δ_Γ^new = SchedulabilityDegree(Γ, ρ)
14         if δ_Γ^new < δ_Γ then -- the schedulability has improved
15             -- offsets are recalculated using ρ^new
16             φ^new = ListScheduling(Γ, M, ψ ρ^new)
17             break -- exit the for-each loop
18         else -- the schedulability has not improved
19             φ.O_i = O_i^old-- restore the old offset
20         end if
21     end for
22 until δ_Γ has not changed or a limit is reached
23 return ρ, φ, δ_Γ
end MultiClusterScheduling
```

Figure 8.2: The MultiClusterScheduling Algorithm

iteration 9–22 continues with a new set of offsets, determined by ListScheduling at line 16, based on the worst-case response times ρ^{new} corresponding to the previously accepted offset.

In the multi-cluster scheduling algorithm, the calculation of offsets is performed by the list scheduling algorithm presented in Figure 8.3. In each iteration, the algorithm visits the processes and messages in the ReadyList. A process or a message in the application is placed in the ReadyList if all its predecessors have been already scheduled. The list is ordered based on the priorities presented in Section 4.3.2. The algorithm terminats when all processes and messages have been visited.

In each loop iteration, the algorithm calculates the earliest time moment (denoted with the variable *offset*) when the process or message $node_i$ in the application graph Γ can start (lines 5–7). There are four situations:

1. The visited node in the application graph is an ET message. In this case, the offset of message m_i is updated to *offset*.
2. The node is a TT message. In this case, the message is sched-

```
ListScheduling(Γ, M, ψ ρ) -- determines the set of offsets φ
1  ReadyList = source nodes of all process graphs in the application
2  while ReadyList ≠ ∅ do
3      node_i = Head(ReadyList)
4      offset = 0 -- determine the earliest time when an activity can start
5      for each direct predecessor node_j of node_i do
6          offset = max(offset, O_j + r_j)
7      end for
8      if node_i is a message m_i then
9          if m_i is an ET message then
10             O_i =offset -- update the message offset
11         else -- m_i is a TT message
12             <round, slot> = ScheduleMessage(offset, s_{m_i}, M(S(m_i)))
13             -- set the TT message offset based on the round and slot
14             O_i = round * T_TDMA + O_slot
15         endif
16     endif
17     else -- node_i is a process P_i
18         if M(P_i) ∈ N_E then -- process P_i is mapped on the ETC
19             O_i = offset -- the ETC process can start immediately
20         else -- process P_i is mapped on the TTC
21             -- P_i has to wait for processor M(P_i) to become available
22             O_i = max(offset, ProcessorAvailable(M(P_i)))
23         end if
24     end if
25     Update(ReadyList)
26 end while
27 return offsets φ
end ListScheduling
```

Figure 8.3: ListScheduling Algorithm

uled using the ScheduleMessage function from Section 4.3.1, which returns the *round* and the *slot* where the frame has been placed (line 12 in Figure 8.3). Once the message has been scheduled, we can determine its offset and worst-case response time (Figure 8.3, line 14). Thus, the offset is equal to the start of the slot in the TDMA round, and the worst-case response time is the slot length.

3. The algorithm visits a process P_i mapped on an ETC node. A process on the ETC can start as soon as its predecessors have finished and its inputs have arrived, hence O_i = *offset* (line 19). However, P_i might, later on, experience interference from higher priority processes.
4. Process P_i is mapped on a TTC node. In this case, besides waiting for the predecessors to finish executing, P_i will also have to wait for its processor $M(P_i)$ to become available (line 22). The earliest time when the processor is available is returned by the ProcessorAvailable function.

Let us now turn the attention back to the multi-cluster scheduling algorithm in Figure 8.2. The algorithm stops when the δ_Γ of the application Γ is no longer improved, or when a limit imposed on the number of iterations has been reached. Since in a loop iteration we do not accept a solution with a larger δ_Γ, the algorithm will terminate when in a loop iteration we are no longer able to improve δ_Γ by modifying the offsets.

8.2.1 SCHEDULABILITY AND RESOURCE ANALYSIS

The analysis in this section is used in the ResponseTimeAnalysis function in order to determine the response times for processes and messages on the ETC. It receives as input the application Γ, the offsets ϕ and the priorities π, and it produces the set ρ of worst case response times.

We have used the response time analysis outlined in Section 6.4.1 for the CAN bus (Equations 6.6, 6.9, 6.10, and 6.11). However, the worst-case queuing delay for a message

(Equation 6.9) is calculated differently depending on the type of message passing employed:

1. From an ETC node to another ETC node (in which case $W_m^{N_i}$ represents the worst-case time a message m has to spend in the Out_{N_i} queue on ETC node N_i). An example of such a message is m_3 in Figure 8.1, which is sent from the ETC node N_3 to the gateway node N_G.
2. From a TTC node to an ETC node (W_m^{CAN} is the worst-case time a message m has to spend in the Out_{CAN} queue). In Figure 8.1, message m_1 is sent from the TTC node N_1 to the ETC node N_2.
3. From an ETC node to a TTC node (where W_m^{TTP} captures the time m has to spend in the Out_{TTP} queue). Such a message passing happens in Figure 8.1, where message m_3 is sent from the ETC node N_3 to the TTC node N_1 through the gateway node N_G where it has to wait for a time W_m^{TTP} in the Out_{TTP} queue.

The messages sent from a TTC node to another TTC node are taken into account when determining the offsets (ListScheduling, Figure 8.2), and thus are not involved directly in the ETC analysis.

The next sections show how the worst-queuing delays and the bounds on the queue sizes are calculated for each of the previous three cases.

From ETC to ETC and from TTC to ETC

The analyses for $W_m^{N_i}$ and W_m^{CAN} are similar. Once m is the highest priority message in the Out_{CAN} queue, it will be sent by the gateway's CAN controller as a regular CAN message, therefore the same equation for W_m can be used:

$$W_m(q) = w_m(q) - qT_m \qquad (8.1)$$

where q is the number of busy periods being examined, and $w_m(q)$ is the width of the level-m busy period starting at time qT_m:

$$w_m(q) = B_m + \sum_{\forall m_j \in hp(m)} \left\lceil \frac{w_m(q) + J_j}{T_j} \right\rceil C_j . \qquad (8.2)$$

The intuition is that m has to wait, in the worst case, first for the largest lower priority message that is just being transmitted (B_m) as well as for the higher priority $m_j \in hp(m)$ messages that have to be transmitted ahead of m (the second term). In the worst case, the time it takes for the largest lower priority message $m_k \in lp(m)$ to be transmitted to its destination is:

$$B_m = \max_{\forall m_k \in lp(m)} (C_k) . \qquad (8.3)$$

Note that in our case, $lp(m)$ and $hp(m)$ also include messages produced by the gateway node, transferred from the TTC to the ETC.

We are also interested to bound the size s_m^{CAN} of the Out_{CAN} and $s_m^{N_i}$ of the Out_{N_i} queue. In the worst case, message m, and all the messages with higher priority than m will be in the queue, awaiting transmission. Summing up their sizes, and finding out what is the most critical instant we get the worst-case queue size:

$$s_{Out} = \max_{\forall m} \left(s_m + \sum_{\forall m_j \in hp(m)} \left\lceil \frac{w_m(q) + J_j}{T_j} \right\rceil C_j \right) \qquad (8.4)$$

where s_m and s_j are the sizes of message m and m_j, respectively.

From ETC to TTC

The time a message m has to spend in the Out_{TTP} queue in the worst case depends on the total size of messages queued ahead of m (Out_{TTP} is a FIFO queue), the size S_G of the gateway slot responsible for carrying the CAN messages on the TTP bus, and the frequency T_{TDMA} with which this slot S_G is circulating on the bus, and thus, the width of the level-m busy period starting at time qT_m is:

$$w_m^{TTP}(q) = B_m + \left\lfloor \frac{(q+1)S_m + I_m(w_m(q))}{S_G} \right\rfloor T_{TDMA}, \quad (8.5)$$

where I_m is the total size of the messages queued ahead of m. Those messages $m_j \in hp(m)$ are ahead of m, which have been sent from the ETC to the TTC, and have higher priority than m:

$$I_f(w_m(q)) = \sum_{\forall m_j \in hp(f)} \left\lceil \frac{w_m(q) + J_j}{T_j} \right\rceil S_j \quad (8.6)$$

where the message jitter J_m is in the worst case the response time of the sender process, $J_m = r_{S(m)}$.

The blocking term B_m is the time interval in which m cannot be transmitted because the slot S_G of the TDMA round has not arrived yet. In the worst case (i.e., the message m has just missed the slot S_G), the frame has to wait an entire round T_{TDMA} for the slot S_G in the next TDMA round.

Determining the size of the queue needed to accommodate the worst case burst of messages sent from the CAN cluster is done by finding out the worst instant of the following sum:

$$s_{Out}^{TTP} = \max_{\forall m}(S_m + I_m). \quad (8.7)$$

8.3 Scheduling and Optimization Strategy

Once we have a technique to determine if a system is schedulable, we can concentrate on optimizing the total queue sizes. Our problem is to synthesize a system configuration ψ such that the application is schedulable, i.e., the condition[1]

$$r_{G_j} \le D_{G_j}, \forall G_j \in \Gamma_i, \quad (8.8)$$

holds, and the total queue size s_{total} is minimized[1]:

$$s_{total} = s_{Out}^{CAN} + s_{Out}^{TTP} + \sum_{\forall N_i \in ETC} s_{Out}^{N_i}. \tag{8.9}$$

In the next section, we propose a resource optimization strategy based on a hill-climb heuristic that uses an intelligent set of initial solutions in order to efficiently explore the design space.

8.3.1 SCHEDULING AND BUFFER OPTIMIZATION HEURISTIC

The basic idea of our buffer optimization heuristic is to find, as a first step, a solution with the smallest possible response times, without considering the buffer sizes, in the hope of finding a schedulable system. This is achieved through the OptimizeSchedule function, outlined in Figure 8.4. Then, a *hill-climbing* heuristic [Ree93] iteratively performs moves intended to minimize the total buffer size while keeping the resulted system schedulable.

The OptimizeSchedule function is a greedy approach which determines an ordering of the slots and their lengths, as well as priorities of messages and processes in the ETC, such that the degree of schedulability δ_Γ (see Section 6.6.1) of the application is maximized.

As an initial TTC bus configuration β, OptimizeSchedule assigns in order nodes to the slots and fixes the slot length to the minimal allowed value, which is equal to the length of the largest message generated by a process assigned to N_i, S_i = <N_i,

1. The worst-case response time of a process graph G_i is calculated based on its sink node as $r_{G_i} = O_{sink} + r_{sink}$. If local deadlines are imposed, they will also have to be tested in the schedulability condition.
1. On the TTC, the synchronization between processes and the TDMA bus configuration is solved through the proper synthesis of schedule tables, thus no output queues are needed. Input buffers on both TTC and ETC nodes are local to processes. There is one buffer per input message and each buffer can store one message instance (see explanation to Figure 3.8 on page 60).

$size_{smallest}$> (line 5 in Figure 8.4). Then, the algorithm starts with the first slot (line 8) and tries to find the node which, when transmitting in this slot, will maximize the degree of schedulability δ_Γ (lines 9–37).

Simultaneously with searching for the right node to be assigned to the slot, the algorithm looks for the optimal slot length (lines 14–32). Once a node is selected for the first slot and a slot length fixed ($S_i = S_{best}$, line 36), the algorithm continues with the next slots, trying to assign nodes (and to fix slot lengths) from those nodes which have not yet been assigned.

When calculating the length of a certain slot we consider the feedback from the MultiClusterScheduling algorithm which recommends slot sizes to be tried out. Before starting the actual optimization process for the bus access scheme, a scheduling of the initial solution is performed which generates the recommended slot lengths. We refer the reader to Section 4.4.1 for details concerning the generation of the recommended slot lengths.

In the OptimizeSchedule function the degree of schedulability δ_Γ is calculated based on the response times produced by the MultiClusterScheduling algorithm (line 21). For the priorities used in the response time calculation we use the "heuristic optimized priority assignment" (HOPA) approach (line 16) from [Gut95], where priorities for processes and messages in a distributed real-time system are determined, using knowledge of the factors that influence the timing behavior, such that the degree of schedulability is improved.

The OptimizeSchedule function also records the best solutions in terms of δ_Γ and s_{total} in the seed_solutions list in order to be used as the starting point for the second step of our OptimizeResources heuristic.

In the first step of our buffer size optimization heuristic OptimizeResources, outlined in Figure 8.5, we have tried to obtain a bus configuration that improves the degree of schedulability of the application. Once a schedulable system is obtained, our goal in the second step is to minimize the buffer space. Our design

```
OptimizeSchedule(Γ, M)
1  -- given an application Γ produces the configuration ψ = <φ, β, π>
2  -- leading to the smallest δΓ
3
4  -- start by determining an initial TTC bus configuration β
5  for each slot Si ∈ β do Si = <Ni, size_smallest> end for
6
7  -- find the best allocation of slots, the TDMA slot sequence
8  for each slot Si ∈ β do
9      for each node Nj ∈ TTC do
10         -- allocate Nj tentatively to Si, Ni gets slot Sj
11         Si = <Nj, size_Sj>
12         Sj = <Ni, size_Si>
13         -- determine best size for slot Si
14         for each slot size ∈ recomended_lengths(Si) do
15             -- calculate the priorities according to HOPA heuristic
16             π = HOPA
17             -- determine the offsets φ,
18             -- thus obtaining a complete system configuration ψ
19             Si = <Nj, size>
20             ψ_current = <φ, β, π>
21             φ = MultiClusterScheduling(Γ, M, ψ_current)
22             -- remember the best configuration so far,
23             -- add it to the seed configurations
24             if δΓ(ψ_current) is best so far then
25                 ψ_best = ψ_current
26                 S_best = Si;
27                 add ψ_best to seed_solutions
28             end if
29             determine s_total for ψ_current
30             if s_total is best so far and Γ is schedulable
31             then add ψ_current to seed_solutions end if
32         end for
33     end for
34     -- make binding permanent, use the S_best corresponding to ψ_best
35     if a S_best exists
36     then Si = S_best end if
37 end for
38
39 return ψ_best, δΓ(ψ_best), seed_solutions
end OptimizeSchedule
```

Figure 8.4: The OptimizeSchedule Algorithm

space exploration in the second step of OptimizeResources (lines 12–22) is based on successive design transformations (generating the neighbors of a solution) called *moves*. For our heuristics, we consider the following types of moves:

- moving a process or a message belonging to the TTC inside its [*ASAP*, *ALAP*] interval calculated based on the current values for the offsets and response times;
- swapping the priorities of two messages transmitted on the ETC, or of two processes mapped on the ETC;

```
OptimizeResources(Γ)
1
2   -- Step 1: try to find a schedulable system
3   seed_solutions = OptimizeSchedule(Γ, M)
4   -- if no schedulable configuration has been found,
5   -- modify mapping and/or architecture
6   if Γ is not schedulable for ψbest then
7       modify mapping
8       go to Step 1
9   end if
10
11
12  -- Step 2: try to reduce the resource need, minimize s_total
13  for each ψ in seed_solutions do
14      repeat
15          -- find moves with highest potential to minimize s_total
16          move_set = GenerateNeighbors(ψ)
17          -- select move which minimizes s_total
18          -- and does not result in an un-schedulable system
19          move = SelectMove(move_set)
20          Perform(move)
21      until s_total has not changed or limit reached
22  end for
23
24  return system configuration ψ,  queue sizes
end OptimizeResources
```

Figure 8.5: The OptimizeResources Algorithm

- increasing or decreasing the size of a TDMA slot with a certain value;
- swapping two slots inside a TDMA round.

The second step of the OptimizeResources heuristic starts from the seed solutions (line 13) produced in the previous step, and iteratively preforms moves in order to reduce the total buffer size, s_{total} (Equation 8.9). The heuristic tries to improve on the total queue sizes, without producing un-schedulable systems. The neighbors of the current solution are generated in the GenerateNeighbours function (line 16), and the move with the smallest s_{total} is selected using the SelectMove function (line 19). Finally, the move is performed, and the loop reiterates. The iterative process ends when there is no improvement achieved on s_{total}, or a limit imposed on the number of iterations has been reached (line 21).

The general limitation of a hill-climbing heuristic is that it can get stuck into a local optimum. In order to improve the chances to find good values for s_{total}, the algorithm has to be executed several times, starting with a different initial solution. The intelligence of our OptimizeResources heuristic lies in the selection of the initial solutions, recorded in the seed_solutions list. The list is generated by the OptimizeSchedule function which records the best solutions in terms of δ_Γ and s_{total}.

Seeding the hill climbing heuristic with several solutions of small s_{total} will guarantee that the local optima are quickly found. However, during our experiments, we have observed that another good set of seed solutions are those that have high degree of schedulability δ_Γ. Starting from a highly schedulable system will permit more iterations until the system degrades to an un-schedulable configuration, thus the exploration of the design space is more efficient.

8.4 Experimental Evaluation

For evaluation of our algorithms we first used applications generated for experimental purpose. We considered two-cluster architectures consisting of 2, 4, 6, 8 and 10 nodes, half on the TTC and the other half on the ETC, interconnected by a gateway. Forty processes were assigned to each node, resulting in applications of 80, 160, 240, 320 and 400 processes. Message sizes were randomly chosen between 8 and 32 bytes. Thirty examples were generated for each application dimension, thus a total of 150 applications were used for experimental evaluation. Worst-case execution times and message lengths were assigned randomly using both uniform and exponential distribution. All experiments were run on a SUN Ultra 10.

In order to provide a basis for the evaluation of our heuristics we have developed two simulated annealing (SA) based algorithms (see Appendix A). Both are based on the moves presented in the previous section. The first one, named SA Schedule (SAS), was set to preform moves such that δ_Γ is minimized. The second one, SA Resources (SAR), uses s_{total} as the cost function to be minimized. Very long and expensive runs have been performed with each of the SA algorithms, and the best ever solution produced has been considered as close to the optimum value.

8.4.1 Scheduling and Bus Access Optimization Heuristics

The first experimental result concerns the ability of our heuristics to produce schedulable solutions. We have compared the degree of schedulability δ_Γ obtained from our OptimizeSchedule (OS) heuristic (Figure 8.4) with the near-optimal values obtained by SAS. Figure 8.6 presents the average percentage deviation of the degree of schedulability produced by OS from the near-optimal values obtained with SAS. Together with OS, a straightforward approach (SF) is presented. For SF we considered a TTC bus configuration consisting of a straightforward ascending order of allocation of the nodes to the TDMA slots; the slot lengths were selected to accommodate the largest message sent by the respective node, and the scheduling has been performed by the MultiClusterScheduling algorithm in Figure 8.2.

Figure 8.6 shows that when considering the optimization of the access to the communication channel, and of priorities, the degree of schedulability improves dramatically compared to the straightforward approach. The greedy heuristic OptimizeSchedule performs well for all the dimensions, having run-times which are more than two orders of magnitude smaller than with SAS. In the figure, only the examples where all the algorithms have obtained schedulable systems were presented. The SF approach failed to find a schedulable system in 26 out of the total 150 applications.

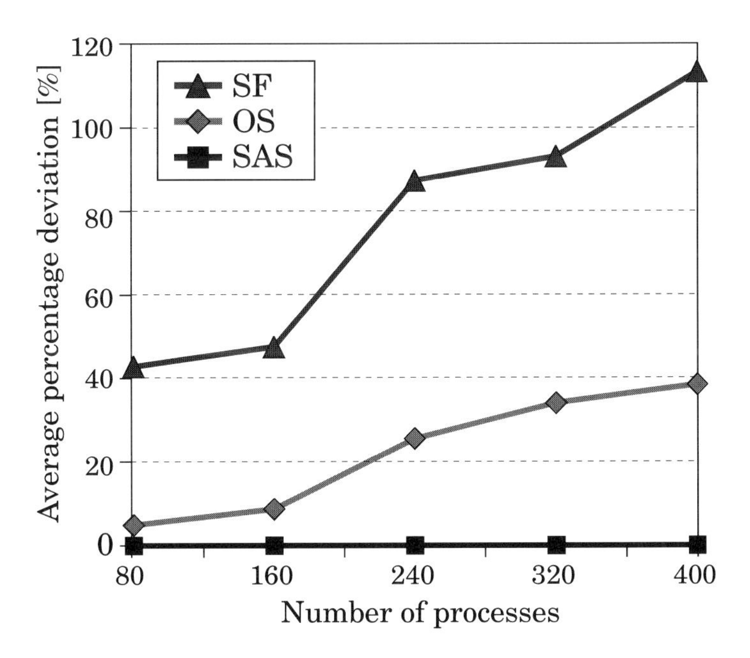

Figure 8.6: Comparison of the Scheduling Optimization Heuristics

8.4.2 BUFFER OPTIMIZATION HEURISTIC

Next, we are interested to evaluate the heuristics for minimizing the buffer sizes needed to run a schedulable application. Thus, we compare the total buffer need s_{total} obtained by the OptimizeResources (OR) function with the near-optimal values obtained when using simulated annealing, this time with the cost function s_{total}. To find out how relevant the buffer optimization problem is, we have compared these results with the s_{total} obtained by the OS approach, which is interested only to obtain a schedulable system, without any other concern. As shown in Figure 8.7a, OR is able to find schedulable systems with a buffer need half of that needed by the solutions produced with OS. The quality of the solutions obtained by OR is also comparable with the one obtained with simulated annealing (SAR).

Another important aspect of our experiments was to determine the difficulty of resource minimization as the number of messages exchanged over the gateway increases. For this, we have generated applications of 160 processes with 10, 20, 30, 40, and 50 messages exchanged between the TTC and ETC clusters. Thirty applications were generated for each number of messages. Figure 8.7b shows the average percentage deviation of the buffer sizes obtained with OR and OS from the near-optimal results obtained by SAR. As the number of inter-cluster messages increases, the problem becomes more complex. The OS approach degrades very fast, in terms of buffer sizes, while OR is able to find good quality results even for intense inter-cluster traffic.

When deciding on which heuristic to use for design space exploration or system synthesis, an important issue is the execution time. In average, our optimization heuristics needed a couple of minutes to produce results, while the simulated annealing approaches (SAS and SAR) had an execution time of up to three hours.

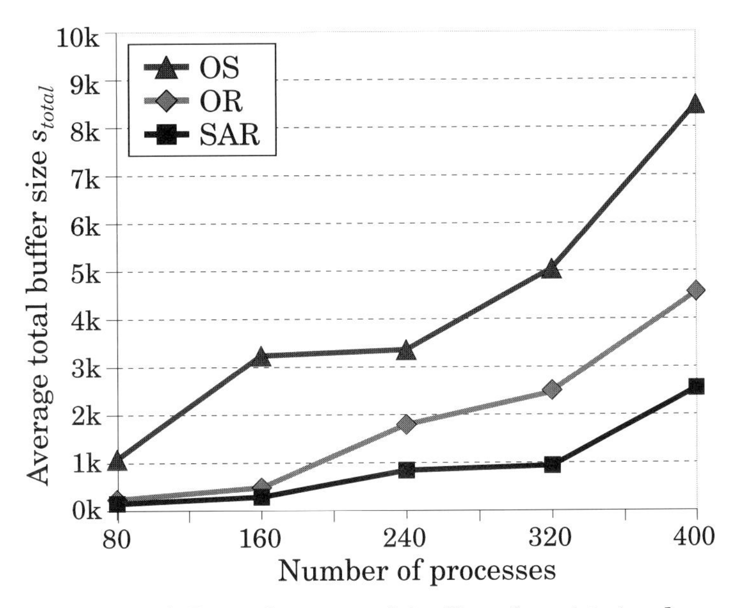

a) Bounds on total buffer size obtained with OS, OR, SAS

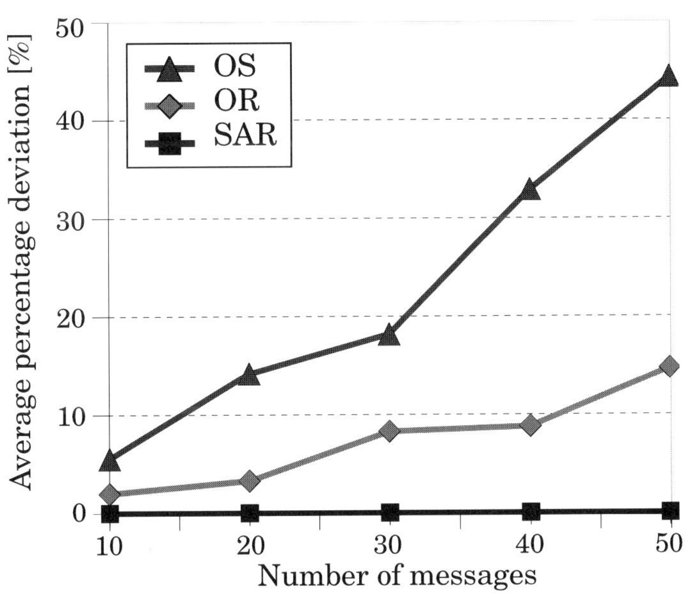

b) Percentage deviations for OS, OR from SAR

Figure 8.7: Comparison of the Buffer Size Minimization Heuristics

8.4.3 THE VEHICLE CRUISE CONTROLLER

Finally, we considered a real-life example implementing a vehicle cruise controller introduced in Section 2.3.3:

- The conditional process graph that models the cruise controller has 32 processes, and is presented in Figure 2.9 on page 40,
- and it was mapped on an architecture consisting of a TTC and an ETC, each with 2 nodes, interconnected by a gateway, as in Figure 2.7b on page 37.
- The software architecture for multi-cluster systems, used by the CC, is presented in Section 3.5.
- We considered one mode of operation with a deadline of 250 ms.

The straightforward approach SF produced an end-to-end response time of 320 ms, greater than the deadline, while both the OS and SAS heuristics produced a schedulable system with a worst-case response time of 185 ms. The total buffer need of the solution determined by OS was 1020 bytes. After optimization with OR a still schedulable solution with a buffer need reduced by 24% has been generated, which is only 6% worse than the solution produced with SAR.

As a conclusion, the optimization heuristics proposed are able to increase the schedulability of the applications and reduce the buffer size needed to run a schedulable application.

In this chapter, the main contribution was the development of a schedulability analysis for multi-cluster systems. However, in the case of both TTP and CAN protocols, several messages share one frame, in the hope to utilize resources more efficiently. Therefore, in the next chapter we propose optimization heuristics for determining frame packing configurations that are able to reduce the cost of the resources needed to run a schedulable application.

Chapter 9
Partitioning and Mapping for Multi-Cluster Systems

USING THE ANALYSIS proposed in the previous chapter, several design optimization problems can be addressed. In the remaining part of the book we will address problems which are characteristic to applications distributed across multi-cluster systems consisting of heterogeneous TT and ET networks. In this chapter, we are interested in the partitioning of the processes of an application into time-triggered and event-triggered domains, and their mapping to the nodes of the clusters. The goal is to produce an implementation which meets all the timing constraints of the application.

The next section presents the design optimization problems we are addressing in this chapter, and Section 9.2 presents our proposed heuristics for the design optimization of multi-cluster systems. The last section of the chapter presents the evaluation of the heuristics.

9.1 Partitioning and Mapping

In this chapter, by partitioning we denote the decision whether a certain process should be assigned to the TT or the ET domain (and, implicitly, to a TTC or an ETC, respectively). Mapping a process means assigning it to a particular node inside a cluster.

Very often, the partitioning decision is taken based on the experience and preferences of the designer, considering aspects like the functionality implemented by the process, the hardness of the constraints, sensitivity to jitter, legacy constraints, etc. Let $\mathcal{P}$ be the set of processes in the application Γ. We denote with $\mathcal{P}_T \subseteq \mathcal{P}$ the subset of processes which the designer has assigned to the TT cluster, while $\mathcal{P}_E \subseteq \mathcal{P}$ contains processes which are assigned to the ET cluster.

Many processes, however, do not exhibit certain particular features or requirements which obviously lead to their implementation as TT or ET activities. The subset $\mathcal{P}^+ = \mathcal{P} \setminus (\mathcal{P}_T \cup \mathcal{P}_E)$ of processes could be assigned to any of the TT or ET domains. Decisions concerning the partitioning of this set of activities can lead to various trade-offs concerning, for example, the schedulability properties of the system, the amount of communication exchanged through the gateway, the size of the schedule tables, etc.

For part of the partitioned processes, the designer might have already decided their mapping. For example, certain processes, due to constraints like having to be close to sensors/actuators, have to be physically located in a particular hardware unit. They represent the sets $\mathcal{P}_T^M \subseteq \mathcal{P}_T$ and $\mathcal{P}_E^M \subseteq \mathcal{P}_E$ of already mapped TT and ET processes, respectively. Consequently, we denote with $\mathcal{P}_T^* = \mathcal{P}_T \setminus \mathcal{P}_T^M$ the TT processes for which the mapping has not yet been decided, and similarly, with $\mathcal{P}_E^* = \mathcal{P}_E \setminus \mathcal{P}_E^M$ the unmapped ET processes. The set $\mathcal{P}^* = \mathcal{P}_T^* \cup \mathcal{P}_E^* \cup \mathcal{P}^+$ then represents all the unmapped processes in the application.

The mapping of messages is decided implicitly by the mapping of processes. Thus, a message exchanged between two processes on the TTC (ETC) will be mapped on the TTP bus (CAN bus) if these processes are allocated to different nodes. If the communication takes place between two clusters, two message instances will be created, one mapped on the TTP bus and one on the CAN bus. The first message is sent from the sender node to the gateway, while the second message is sent from the gateway to the receiving node.

Example 9.1: Let us illustrate some of the issues related to partitioning in such a context. In the example presented in Figure 9.1 we have an application[1] with six processes, P_1 to P_6, and four nodes, N_1 and N_2 on the TTC, N_3 on the ETC and the gateway node N_G. The worst-case execution times on each node are given to the right of the application graph. Note that N_2 is faster than N_3, and an "X" in the table means that the process is not allowed to be mapped on that node. The mapping of P_1 is fixed on N_1, P_3 and P_6 are mapped on N_2, P_2 and P_5 are fixed on N_3, and we have to decide how to partition P_4 between the TT and ET domains. Let us also assume that process P_5 is the highest priority process on N_3. In addition, P_5 and P_6 have each a deadline, D_5 and D_6, respectively, as illustrated in the figure by thick vertical lines.

We can observe that although P_3 and P_4 do not have individual deadlines, their mapping and scheduling has a strong impact on their successors, P_5 and P_6, respectively, which are deadline constrained. Thus, we would like to map P_4 such that not only P_3 can start on time, but P_4 also starts soon enough to allow P_6 to meet its deadline.

As we can see from Figure 9.1a, this is impossible to achieve by mapping P_4 on the TTC node N_2. It is interesting to observe that, if preemption would be allowed in the TT domain, as in Figure 9.1b, both deadlines could be met. This,

1. Communications are ignored for this example.

however, is impossible on the TTC where preemption is not allowed. Both deadlines can be met only if P_4 is mapped on the slower ETC node N_3, as depicted in Figure 9.1c. In this case, although P_4 competes for the processor with P_5, due to the preemption of P_4 by the higher priority P_5, all deadlines are satisfied.

■

For a multi-cluster architecture the communication infrastructure has an important impact on the design and, in particular, the mapping decisions.

Example 9.2: Let us consider the example in Figure 9.2. We assume that P_1 is mapped on node N_1 and P_3 on node N_3 on the TTC, and we are interested to map process P_2. P_2 is allowed to be mapped on the ETC node N_2 or on the ETC node

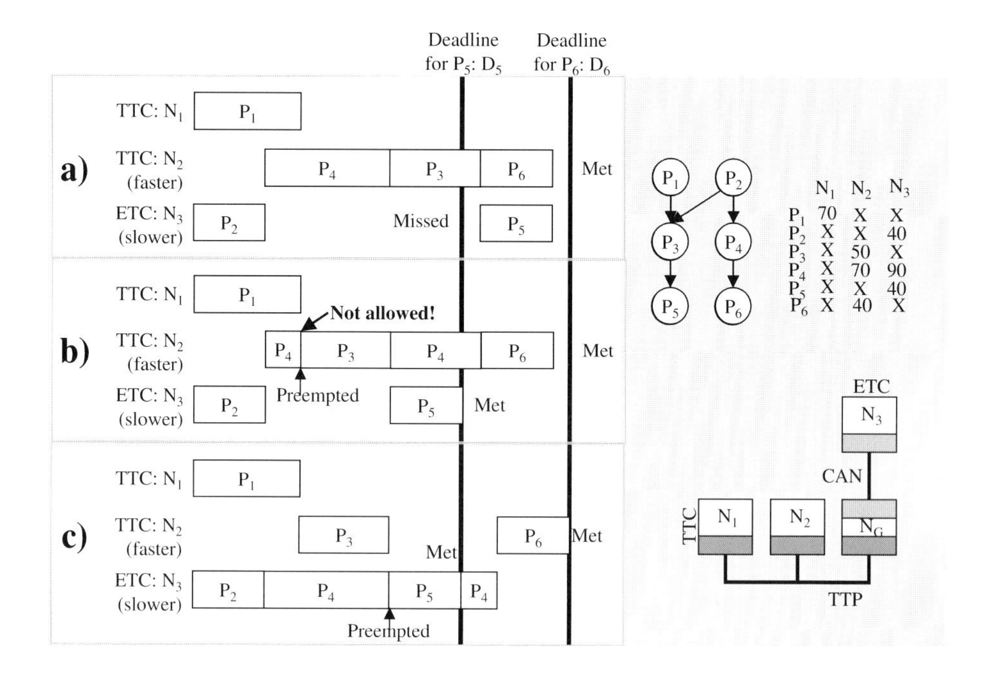

Figure 9.1: Partitioning Example

N_4, and its execution times are depicted in the table to the right of the application graph.

In order to meet the deadline, one would map P_2 on the node it executes fastest, N_2 on the TTC, see Figure 9.2a. However, this will lead to a deadline miss due to the TTP slot configuration which introduces communication delays. The application will meet the deadline only if P_2 is mapped on the slower node, i.e., node N_4 in the case in Figure 9.2b[1]. Not only is N_4 slower than N_2, but mapping P_2 on N_4 will place P_2 on a different cluster than P_1 and P_3, introducing extra communication delays through the gateway node. However, due to the actual communication configuration, the mapping alternative in Figure 9.2b is desirable.

■

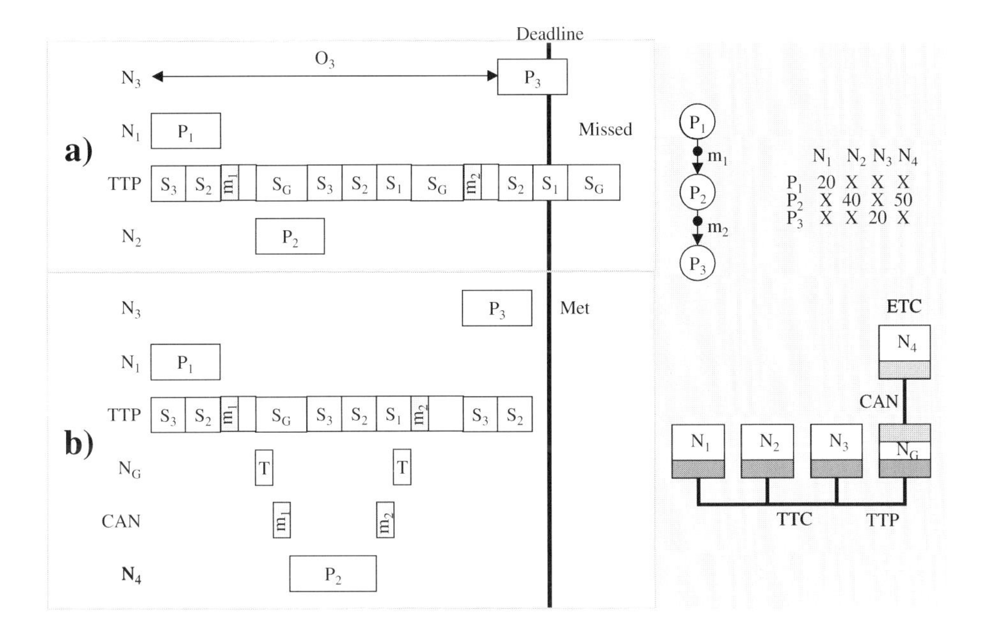

Figure 9.2: Mapping Example

1. Process T in Figure 9.2b executing on the gateway node N_G is responsible for transferring messages between the TTP and CAN controllers.

9.1.1 EXACT PROBLEM FORMULATION

As an input we have an application Γ given as a set of process graphs (Section 2.3.1) and a two-cluster system consisting of a TT and an ET cluster (Section 3.5). As introduced previously, $\mathcal{P}_T$ and $\mathcal{P}_E$ are the sets of processes already partitioned into TT and ET, respectively. Also, $\mathcal{P}_T^M \subseteq \mathcal{P}_T$ and $\mathcal{P}_E^M \subseteq \mathcal{P}_E$ are the sets of already mapped TT and ET processes.

We are interested to find a partitioning for processes in $\mathcal{P}^+ = \mathcal{P} \setminus (\mathcal{P}_T \cup \mathcal{P}_E)$ and decide a mapping for processes in $\mathcal{P}^* = \mathcal{P}_T^* \cup \mathcal{P}_E^* \cup \mathcal{P}^+$, where $\mathcal{P}_T^* = \mathcal{P}_T \setminus \mathcal{P}_T^M$, and $\mathcal{P}_E^* = \mathcal{P}_E \setminus \mathcal{P}_E^M$ such that imposed deadlines are guaranteed to be satisfied.

9.2 Partitioning and Mapping Strategy

The design problem formulated in the previous section is NP-complete. The MultiClusterConfiguration strategy (MCC) we propose to solve the partitioning and mapping problem has three steps:

1. In the first step, we decide very quickly on an initial bus access configuration. The initial bus access configuration is determined, for the TTC, by assigning in order nodes to the slots ($S_i = N_i$) and fixing the slot length to the minimal allowed value, which is equal to the length of the largest message in the application. For the ETC we calculate the message priorities π based on the deadlines of the receiver processes.
2. Once an initial bus access configuration has been decided, in the second step, we decide an initial partitioning and mapping, using the algorithm described in Section 9.2.1. After the initial partitioning and mapping are obtained, the application is scheduled using the MultiClusterScheduling algorithm outlined in the previous chapter.
3. If the application is schedulable the optimization strategy stops. Otherwise, it continues with the third step by using an

iterative improvement heuristic, namely PMHeuristic, presented in Section 9.2.2, to improve the partitioning and mapping obtained in the first step.

After these steps the system can be further optimized by improving the access to the communication infrastructure. Such an optimization step is presented in the next chapter.

9.2.1 INITIAL PARTITIONING AND MAPPING (IPM)

Our initial mapping and partitioning algorithm (InitialPM, Figure 9.3) receives as input the merged graph $\mathcal{G}_\Gamma$ obtained by merging all the graphs of the application Γ (as described in Figure 5.5 on page 111).

The IPM algorithm uses a list scheduling based greedy approach. A process P_i is placed in the ready list $\mathcal{L}$ if all its predecessors have been already scheduled. In each iteration of the loop (lines 2–7), all ready processes from the list $\mathcal{L}$ are investigated, and that process P_i is selected for mapping by the SelectProcess function, which has the largest delay $\delta_i = r_i + l_i$. In the previous equation, r_i is the response time of process P_i on the fastest node in $\mathcal{N}_{P_i}$, and l_i is the critical path starting from process P_i, defined as:

```
InitialPM(𝒢Γ) -- Initial Partitioning and Mapping
1  ℒ = {source of 𝒢Γ} -- start with the first node of the merged graph
2  while ℒ ≠ ∅ do -- visits ready processes in the order of list scheduling
3      P = SelectProcess(ℒ)
4      N = SelectNode(𝒩P)
5      M(P) = N -- map process P on node N
6      ℒ = UpdateReadyList(ℒ)
7  end while
end InitialPM
```

Figure 9.3: The Initial Partitioning and Mapping

$$l_i = \max_k \sum_{\forall \tau_j \in \pi_{ik}} r_{\tau_j} \qquad (9.1)$$

where π_{ik} is the kth path from process P_i to the sink node of $\mathcal{G}_\Gamma$ (not including P_i), and r_{τ_j} is the response time of a process or message on π_{ik}. The response times are calculated using the MultiClusterScheduling function, under the following assumptions:

- Every yet unpartitioned/unmapped process $P_i \in \mathcal{P}^*$ is considered mapped on the fastest node from the list of potential nodes $\mathcal{N}_{P_i}$.
- The worst-case response time for messages sent or received by yet unpartitioned/unmapped processes is considered equal to zero.

Example 9.3: Let us consider the design example in Figure 9.4 where we have five processes, P_1 to P_5, and three nodes, N_1 on the TTC, N_2 on the ETC and the gateway node N_G. The initial bus configuration, consisting of the slots order and size, together with the ET message priorities, is also given. The mapping of P_3 is fixed on N_1, P_5 is fixed on N_2, and we have to decide where to partition and map P_1, P_2 and P_4. In the first iteration of IPM, SelectProcess has to decide between P_1 and P_2 which are ready for execution. The critical path of P_1 is $l_1 = \max(r_{m_1} + r_3 + r_{m_4} + r_5, r_{m_2} + r_4 + r_{m_5} + r_5) = \max(0 + 40 + 40 + 40, 0 + 30 + 0 + 40) = 120$, while $l_2 = r_{m_3} + r_4 + r_{m_5} + r_5 = 0 + 30 + 0 + 40 = 70$. Thus, the delay of P_1 is $\delta_1 = C_1^{N_2} + l_1 = 30 + 120 = 150$, and the delay of P_2 is $\delta_2 = C_2^{N_2} + l_2 = 60 + 70 = 130$[1]. Therefore, SelectProcess will select P_1 because it has a larger delay.

■

Once a process P_i is selected, all mapping alternatives of P_i to nodes[2] in $\mathcal{N}_{P_i}$ are tested by the SelectNode function. Out of these

1. According to the first assumption, both P_1 and P_2 are considered mapped on the fastest node.

alternatives, SelectNode returns that node N_k which leads to the smallest end-to-end delay $\delta_i^{N_k}$ on the application graph:

$$\delta_i^{N_k} = O_i^{N_k} + r_i^{N_k} + l_i^{N_k}. \tag{9.2}$$

In the previous equation, $O_i^{N_k}$ is the offset of process P_i when mapped on node N_k (i.e., the earliest possible starting time taking into account the predecessors and the communication delay

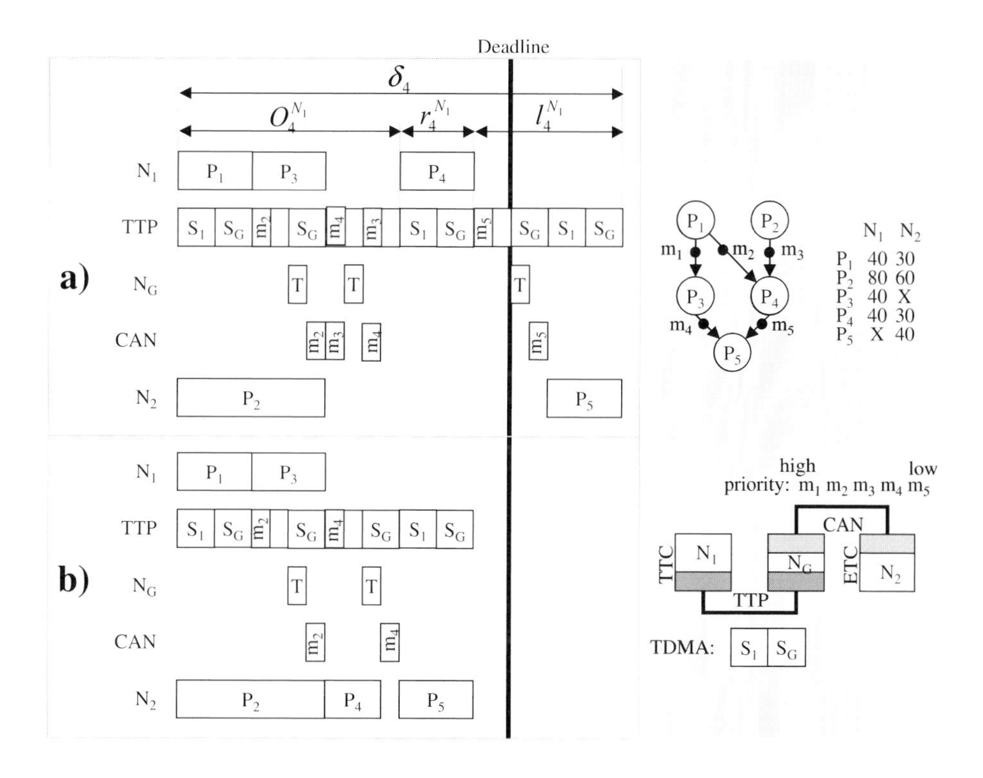

Figure 9.4: Design Example

2. $\mathcal{N}_{P_i}$ is the set of nodes on which process P_i could, potentially, be executed. If the process is already partitioned to a certain cluster, only nodes in that cluster are considered.

of the incoming messages) calculated by our scheduling algorithm.

The worst-case response time $r_i^{N_k}$ is equal to the worst-case execution time $C_i^{N_k}$ if N_k is in the TTC ($N_k \in \mathcal{N}_T$). If N_k is in the ETC ($N_k \in \mathcal{N}_E$), the worst-case response time is calculated according to the Equations 6.4 and 6.5.

The third term of the delay $\delta_i^{N_k}$ represents the critical path from process P_i to the sink node (as introduced in Equation 9.1) in the case P_i is mapped on node N_k. The delay $\delta_i^{N_k}$ is calculated by the MultiClusterScheduling function, under the same assumptions mentioned earlier.

Example 9.4: Let us go back to the example in Figure 9.4. IPM has decided in the first two iterations of the while loop (lines 2–7 in Figure 9.3) that P_1 should be mapped on N_1 and P_2 on N_2. In the third iteration, P_4 has been selected by SelectProcess, and now the mapping alternatives on N_1 and N_2 are tested by the SelectNode function. According to Equation 9.2, if P_4 is mapped on N_1 we have $\delta_4^{N_1} = O_4^{N_1} + r_4^{N_1} + l_4^{N_1} = 120 + 40 + (40 + 40) = 240$ (see Figure 9.4a). Similarly, for the alternative on N_2 we have $\delta_4^{N_2} = O_4^{N_2} + r_4^{N_2} + l_4^{N_2} = 80 + 30 + (0 + 40) = 150$ (see Figure 9.4b). Thus, P_4 will be mapped on N_2 which produces the smallest delay of 150. IPM will finally produce the schedulable solution presented in Figure 9.4b. ■

9.2.2 PARTITIONING AND MAPPING HEURISTIC (PMH)

If, after the initial partitioning, mapping and bus setup we do not obtain a schedulable application, we apply an iterative improvement algorithm, the PMHeuristic in Figure 9.5. The algorithm receives as input the application Γ, the initial partitioning and mapping M^0 produced by IPM and produces a partitioning and mapping for processes in $\mathcal{P}^*$.

We investigate each unschedulable graph $G_i \in \Gamma$, i.e., the response time r_{G_i} is larger than the deadline D_{G_i}. Our heuristic is to perform changes to the mapping of processes in Γ that would reduce the critical path of G_i, and thus the worst-case response time r_{G_i}.

In each iteration, the algorithm selects that unschedulable process graph G_i which has the maximum delay $\Delta_{G_i} = r_{G_i} - D_{G_i}$ between its response time and the deadline (line 2). Let us denote the maximum delay with Δ_{max}, and the corresponding graph with G_{max}. Next, we determine the critical path $\mathcal{P}_{CP}$ of the process graph G_{max}. For example, for the process graph in

```
PMHeuristic(Γ, M) -- Partitioning and Mapping Heuristic
1  while (∃ G_i ∈ Γ ∧ r_Gi > D_Gi) and (Δ_max improved in the previous iteration) do
2      Δ_max = maximum of r_Gi – D_Gi, ∀ G_i ∈ Γ ∧ r_Gi > D_Gi
3      G_max = graph corresponding to Δ_max
4      P_CP = FindCriticalPath(G_max)
5      for each P_i ∈ P_CP do -- find changes with a potential to improve r_Gmax
6          if M(P_i) ∈ N_T then
7              List = ProposedTTChanges(P_i)
8          else -- in this case M(P_i) ∈ N_E
9              List = ProposedETChanges(P_i)
10         end if
11         for each ProposedChange ∈ List do -- determine the improvement
12             Perform(ProposedChange); MultiClusterScheduling(Γ, M)
13             Δ_max = maximum of r_Gi – D_Gi, ∀ G_i ∈ Γ ∧ r_Gi > D_Gi
14             if Δ_max smallest so far then
15                 BestChange = ProposedChange
16             end if
17             Undo(ProposedChange)
18         end for
19         -- apply the move improving the most
20         If ∃ BestChange then Perform(BestChange) end if
21     end for
22 end while
23 return M
end PMHeuristic
```

Figure 9.5: The Partitioning and Mapping Heuristic

Figure 9.4 scheduled as in case (a), the critical path is composed of P_2, P_4 and P_5.

The intelligence of the heuristic lies in how it determines changes (i.e., design transformations) to the mapping of processes that potentially can lead to a shortening of the critical path (lines 7 and 9). The list of proposed changes *List* leading to a potential improvement are then evaluated (lines 11–18) to find out the change that produces the largest reduction of Δ_{max}, which is finally applied to the system configuration (line 20). Reducing Δ_{max} means, implicitly, reducing the response time of the process graph G_{max} investigated in the current iteration. The algorithm terminates if all graphs in the application are schedulable, or no improvement to Δ_{max} is found.

Since a call to MultiClusterScheduling that evaluates the changes is costly in terms of execution time, it is crucially to find out a short list of proposed changes that will potentially lead to the largest improvement. Looking at Equation 9.2, we can observe that the length of the critical path $\mathcal{P}_{CP}$ would be reduced if, for a process $P_i \in \mathcal{P}_{CP}$, we would:

1. reduce the offset O_i (first term of Equation 9.2);
2. decrease the worst-case response time r_i (second term);
3. reduce the critical path from P_i to the sink node (third term).

To reduce (1) we have to reduce the delay of the communication from P_i's predecessors to P_i. Thus, we consider transformations that would change the mapping of process P_i and of predecessors of P_i such that the communication delay is minimized. However, only those predecessors are considered for remapping which actually delay the execution of P_i. Let us go back to Figure 9.4, and consider that PMH starts from an initial partitioning and mapping as depicted in Figure 9.4a. In this case, to reduce the offset O_4 of process P_4, we will consider mapping P_4 on node N_2 as depicted in Figure 9.4b, reducing thus the offset from 120 to 80.

The approach to reduce (2) depends on the type of process. Both for TT and ET processes we can decrease the worst-case execution time C_i by selecting a faster node. For example, in Figure 9.4, by moving P_4 from N_2 to N_1 we reduce its worst-case execution time from 40 to 30. However, for ET processes we can further reduce r_i by investigating the interference from other processes on P_i (Equation 6.5). Thus, we consider mapping processes with a priority higher than P_i on other nodes, reducing thus the interference.

Point (3) is concerned with the critical path from process P_i to the sink node. In this case, we are interested to reduce the delay of the communication from P_i to its successor process on the critical path. This is achieved by considering changes to the mapping of P_i or to the mapping of the successor process (e.g., by including them in the same cluster, same processor, etc.). For example, in Figure 9.4a, the critical path of P_4 is enlarged by the communication delay due to m_5 exchanged by P_4 on the TTC with P_5 on the ETC. To reduce the length of the critical path we will consider mapping P_4 to N_2, and thus the communication will take place on the same processor.

9.3 Experimental Evaluation

For the evaluation of our algorithms we used applications of 50, 100, 150, 200, and 250 processes (all unpartitioned and unmapped), to be implemented on two-cluster architectures consisting of 2, 4, 6, 8, and 10 different nodes, respectively, half on the TTC and the other half on the ETC, interconnected by a gateway.

Thirty examples were randomly generated for each application dimension, thus a total of 150 applications were used for experimental evaluation. We generated both graphs with random structure and graphs based on more regular structures like trees and groups of chains. Execution times and message

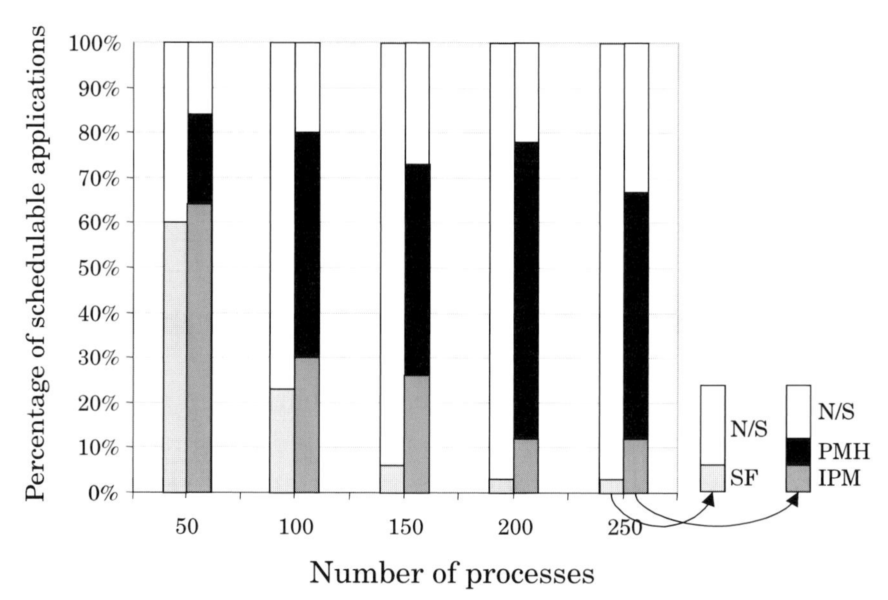

a) Percentace of schedulable applications

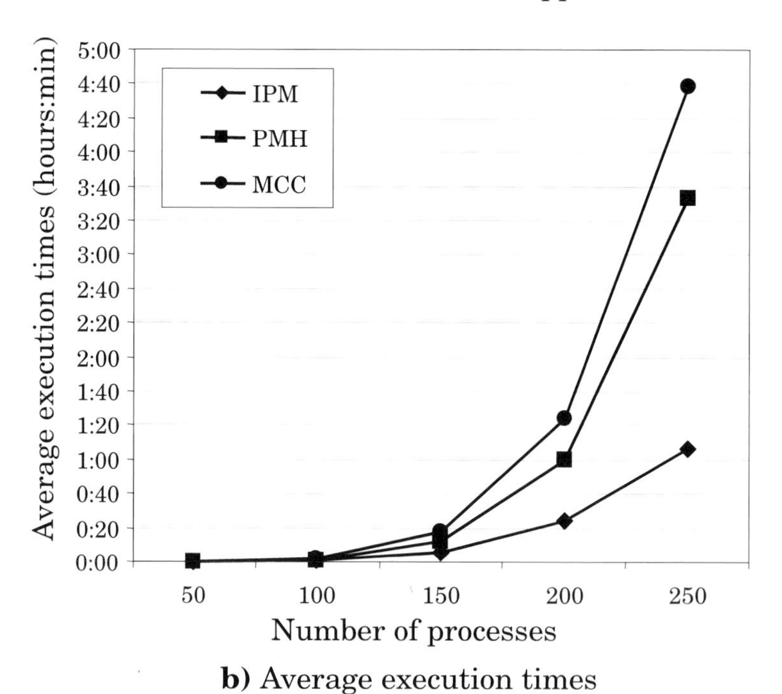

b) Average execution times

Figure 9.6: Comparison of the Partitioning and Mapping Heuristics

lengths were assigned randomly using both uniform and exponential distribution within the 10 to 100 ms, and 2 to 8 bytes ranges, respectively. The experiments were done on SUN Ultra 10 computers.

We were interested to evaluate the proposed approaches. Hence, we have implemented each application, on its corresponding architecture, using the MultiClusterConfiguration (MCC) strategy presented in Section 9.2. Figure 9.6a presents the number of schedulable solutions found after each step of our optimization strategy (N/S stands for "not schedulable"). Together with the MCC steps, Figure 9.6a also presents a straightforward solution (SF). The SF approach performs a partitioning and mapping that tries to balance the utilization among nodes and buses. This is a configuration which, in principle, could be elaborated by a careful designer without the aid of optimization tools like the one proposed here.

Out of the total number of applications, only 19% were schedulable with the implementation produced by SF. However, using our MCC strategy, we are able to obtain schedulable applications in 76% of the cases: 30% after step two (IPM), and 76% after step three (PMH). It is easy to observe that, for all application dimensions, by performing the proposed optimization steps, large improvements over the straightforward configuration could be produced. Moreover, as the applications become larger, it is more difficult for SF to find schedulable solutions, while the optimization steps of MCC perform very well. For 150 processes, for example, MCC has been able to find schedulable implementations for 83% of the applications. The bottom bar, corresponding for 26%, is the percentage of schedulable applications found by IPM. On top of that, PMH, depicted by a black bar, adds another 47%.

Figure 9b presents the execution times for the IPM and PMH steps of our multi-cluster configuration strategy, as well as for the complete algorithm (MCC). Note that the times presented in the figure for MCC include a complete optimization loop, that performs partitioning, mapping, bus access optimization and scheduling. The complete optimization process implemented by the MCC strategy takes under five hours for very large process graphs of 250 processes, while for applications consisting of 100 processes it takes on average 2 minutes.

9.3.1 THE VEHICLE CRUISE CONTROLLER

Finally, we considered the cruise controller example presented in Section 2.3.3:

- The model for the cruise controller is presented in Figure 2.9 on page 40.
- The architecture, consisting of a TTC and an ETC, each with 2 nodes, interconnected by the CEM node, is depicted in Figure 2.7b on page 37.
- The software architecture for multi-cluster systems, used by the CC, is presented in Section 3.5.
- We have considered a deadline of 150 ms.

In this setting, the SF approach failed to produce a schedulable implementation, leading to response time of 392 ms. After IPM (second step of MCC), we were able to reduce the response time of the MCC to 154, which is still larger than the deadline. However, applying PMH (step three) we are able to obtain a schedulable implementation with a response time of 146 ms. The ceomplete MCC executes for under two minutes for the CC.

The evaluation using synthetic applications, as well as the real-life example, show that by using our optimization approaches for the partitioning and mapping problem, we are able to find schedulable implementations under limited resources, achieving an efficient utilization of the system.

Chapter 10
Schedulability-Driven Frame Packing for Multi-Cluster Systems

THE PREVIOUS CHAPTERS have presented analysis methods for communication-intensive heterogeneous real-time systems, taking into account the details of the communication protocols, in our case CAN and TTP.

We have, however, not addressed the issue of *frame packing*, which is of utmost importance in cost-sensitive embedded systems where resources, such as communication bandwidth, have to be fully utilized [Kop95], [San00], [Raj98]. In both TTP and CAN protocols messages are not sent independently, but several messages having similar timing properties are usually packed into frames. In many application areas like, for example, automotive electronics, messages range from one single bit (e.g., the state of a device) to a couple of bytes (e.g., vehicle speed, etc.). Transmitting such small messages one per frame would create a high communication overhead, which can cause long delays leading to an unschedulable system. For example, 65 bits have to be transmitted on CAN for delivering one single bit of application data. Moreover, a given frame configuration defines the exact behavior of a node on the network, which is very important when integrating nodes from different suppliers.

The issue of frame packing (sometimes referred to as frame compiling) has been previously addressed separately for the CAN and the TTP. In [San00], [Raj98] CAN frames are created based on the properties of the messages, while in [Kop95] a "cluster compiler" is used to derive the frames for a TT system which uses TTP as the communication protocol. However, researchers have not addressed frame packing on multi-cluster systems implemented using both ET and TT clusters, where the interaction between the ET and TT processes of a hard real-time application has to be very carefully considered in order to guarantee the timing constraints. As our multi-cluster scheduling strategy in Section 8.2 shows, the issue of frame packing cannot be addressed separately for each type of cluster, since the inter-cluster communication creates a circular dependency.

Therefore, in this chapter, we concentrate on the issue of packing messages into frames, for multi-cluster distributed embedded systems consisting of time-triggered and event-triggered clusters, interconnected via gateways. We are interested to obtain that frame configuration which would produce a schedulable system. We have updated our schedulability analysis presented in Section 8.2 to account for the frame packing, and we have proposed two optimization heuristics that use the schedulability analysis as a driver towards a frame configuration that leads to a schedulable system.

The chapter is organized in three sections. The next section presents the exact formulation of the problem that we are addressing in this chapter. Section 10.2 updates the schedulability analysis for multi-clusters developed in the previous chapter, and uses it to drive the optimization heuristics used for frame generation. The last section presents the experimental results.

10.1 Problem Formulation

As input to our problem we have an application Γ given as a set of conditional process graphs mapped on an architecture consisting of a TTC and an ETC interconnected through a gateway.

As part of our frame packing approach, we are interested to generate all the MEDLs on the TTC (i.e., the TT frames and the sequence of the TDMA slots), as well as the ET frames and their priorities on the ETC such that the global system is schedulable.

More formally, we are interested to find a mapping of messages to frames (a frame packing configuration) denoted by a 4-tuple $\psi=<\alpha, \pi, \beta, \sigma>$ such that the application Γ is schedulable. Once a schedulable system is found, we are interested to further improve the degree of schedulability (defined in Section 6.6.1), so the application can potentially be implemented on a cheaper hardware architecture (with slower buses and processors).

Determining a frame configuration ψ means deciding on:

- The mapping of application messages transmitted on the ETC to frames (the set of ETC frames α), and their relative priorities π. Note that the ETC frames α have to include messages transmitted from an ETC node to a TTC node, messages transmitted inside the ETC cluster, and those messages transmitted from the TTC to the ETC.
- The mapping of messages transmitted on the TTC to frames, denoted by the set of TTC frames β and the sequence σ of slots in a TDMA round. The slot sizes are determined based on the set β, and are calculated such that they can accommodate the largest frame sent in that particular slot. We consider that messages transmitted from the ETC to the TTC are not statically allocated to frames. Rather, we will dynamically pack messages originating from the ETC into the "gateway frame," for which we have to decide the data field length.

Example 10.1: Let us consider the example in Figure 10.1, where we have the process graph G in Figure 10.1d mapped

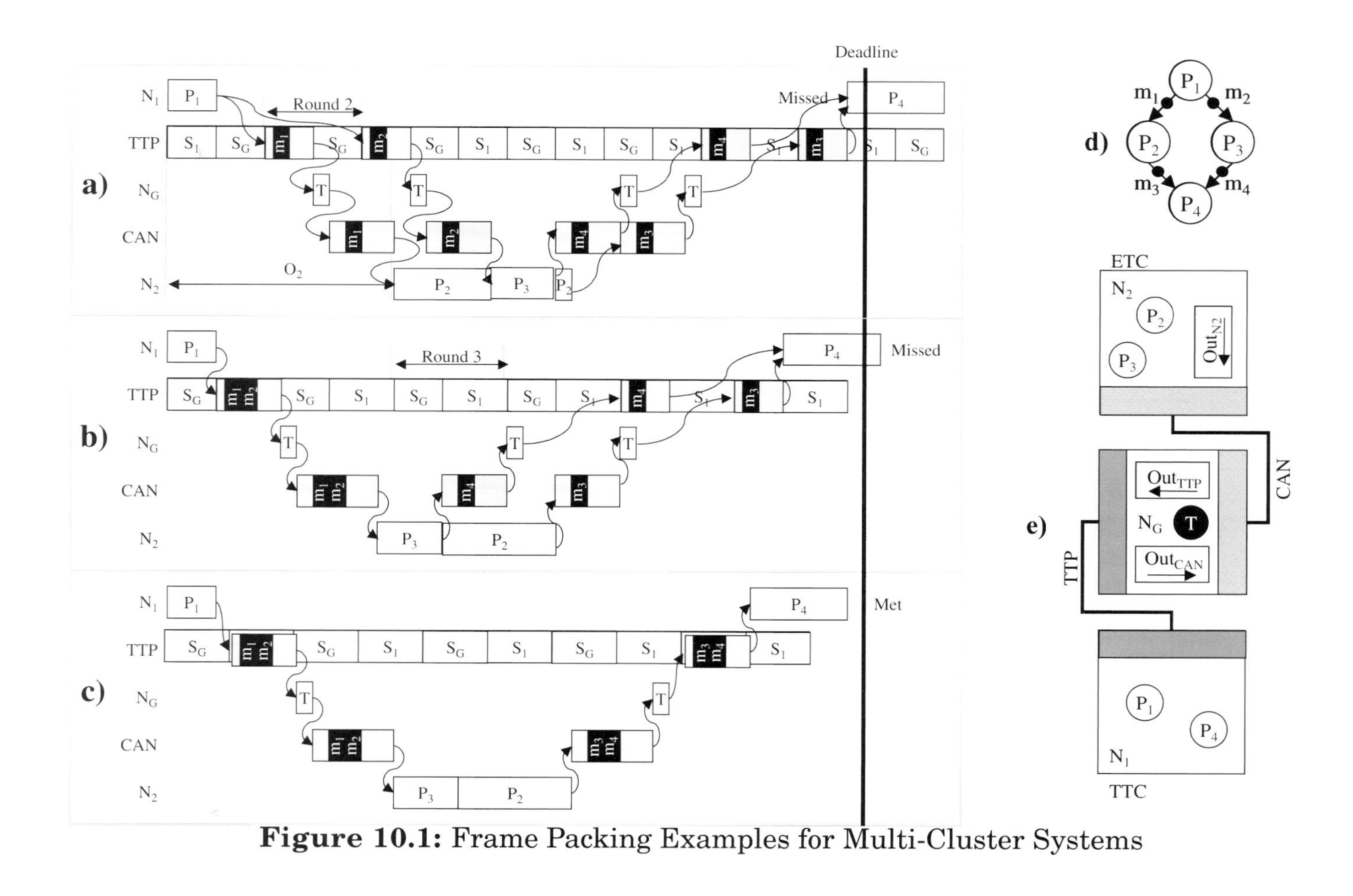

Figure 10.1: Frame Packing Examples for Multi-Cluster Systems

on the two-cluster system as indicated in Figure 10.1e.
In the system configuration of Figure 10.1a we consider that, on the TTP bus, the node N_1 transmits in the first slot (S_1) of the TDMA round, while the gateway transmits in the second slot (S_G). Process P_3 has a higher priority than process P_2, hence P_2 will be interrupted by P_3 when it receives message m_2. In such a setting, P_4 will miss its deadline, which is depicted as a thick vertical line in Figure 10.1. Changing the frame configuration as in Figure 10.1b, so that messages m_1 and m_2 are packed into frame f_1 and slot S_G of the gateway comes first, processes P_2 and P_3 will receive m_1 and m_2 sooner and thus reduce the worst-case response time of the process graph, which is still larger than the deadline. In Figure 10.1c, we also pack m_3 and m_4 into f_2. In such a situation, the sending of m_3 will have to be delayed until m_4 is queued by P_2. Nevertheless, the worst-case response time of the application is further reduced, which means that the deadline is met, thus the system is schedulable.
However, packing more messages will not necessarily reduce the worst-case response times further, as it might increase too much the worst-case response times of messages that have to wait for the frame to be assembled, like is the case with message m_3 in Figure 10.1c. We are interested to find a frame packing that leads to a schedulable system.

■

10.2 Frame Packing Strategy

We have updated the schedulability analysis for an ETC cluster, presented in Section 8.2, to consider frames. We consider that the response time of a message m is equal to the response time of the frame f in which message m is transmitted. The response time of a frame f is calculated similar to a the worst-case response time for a message in Section 8.2, with the following exceptions:

- The size of the frame is calculated taking into account the exact frame configuration for TTP and CAN (see Figures 3.3 and 3.4 on page 49 and page 50, respectively) and the size of the messages packed into the frame.
- The jitter of a frame f is, in the worst case, equal to the largest worst case response time $r_{S(m)}$ of a sender process $P_{S(m)}$ which sends message m packed into frame f:

$$J_f = \max_{\forall m \in f}(r_{S(m)}) . \quad (10.1)$$

For the scheduling of a multi-cluster system we use the same algorithm as in Figure 8.2. Once we have a technique to determine if a system is schedulable, we can concentrate on optimizing the packing of messages to frames.

Such an optimization problem is NP complete [San00], thus obtaining the optimal solution is not feasible. We propose two frame packing optimization strategies, one based on a simulated annealing approach, while the other is based on a greedy heuristic that uses intelligently the problem-specific knowledge in order to explore the design space.

In order to drive our optimization algorithms towards schedulable solutions, we characterize a given frame packing configuration using the degree of schedulability of the application, as presented in Section 6.6.1.

10.2.1 Frame Packing with Simulated Annealing

The first algorithm we have developed is based on a simulated annealing (SA) strategy, see Appendix A. As discussed before, an essential component of an SA algorithm is the generation of a new solution x' starting from the current one x_{now} (Figure A.1 in Appendix A). The neighbors of the current solution x_{now} are obtained by performing transformations (called moves) on the current frame configuration ψ We consider the following moves:

- moving a message m from a frame f_1 to another frame f_2 (or moving m into a separate single-message frame);
- swapping the priorities of two frames in α;
- swapping two slots in the sequence σ of slots in a TDMA round.

10.2.2 FRAME PACKING GREEDY HEURISTIC

The OptimizeFramePacking greedy heuristic (Figure 10.2) constructs the solution by progressively selecting the best candidate in terms of degree of schedulability.

We start by observing that all activities taking place in a multi-cluster system are ordered in time using the offset information, determined in the StaticScheduling function (see Section 8.2) based on the response times known so far and the application structure (i.e., the dependencies in the process graphs). Thus, our greedy heuristic outlined in Figure 10.2, starts with building two lists of messages ordered according to the ascending value of their offsets, one for the TTC, $messages_\beta$, and the other for ETC, $messages_\alpha$. Our heuristic is to consider for packing in the same frame messages which are adjacent in the ordered lists.

Example 10.2: For example, let us consider that we have three messages, m_1 of 1 byte, m_2 of 2 bytes and m_3 of 3 bytes, and that messages are ordered as m_3, m_1, m_2 based on the offset information. Also, assume that our heuristic has suggested two frames, frame f_1 with a data field of 4 bytes, and f_2 with a data field of 2 bytes.

The PackMessages function will start with m_3 and pack it in frame f_1. It continues with m_2, which is also packed into f_1, since there is space left for it. Finally, m_3 is packed in f_2, since there is no space left for it in f_1.

■

```
OptimizeFramePacking(Γ)
1  -- given an application Γ, find out if it is schedulable and produce the
2  -- configuration ψ = <α, π, β, σ> leading to the smallest δ_Γ
3  -- build the message lists ordered ascending on their offsets
4  messages_β = ordered list of n_β messages on the TTC
5  messages_α = ordered list of n_α messages on the ETC
6  -- build an initial frame configuration ψ = <α, π, β, σ>
7  β = messages_β; α = messages_α -- initially, each frame carries one message
8  -- determine an initial TDMA slot sequence σ
9  for each slot S_i ∈ σ do S_i = N_i; size_S_i = size_largest message end for
10 π_initial = HOPA -- calculate the priorities π according to the HOPA heuristic
11 -- find the best allocation of slots, the TDMA slot sequence σ_current
12 for each slot S_i ∈ σ_current do
13   for each node N_j ∈ TTC do
14     σ_current.S_i = N_j; σ_current.S_j = N_i -- allocate N_j tentatively to S_i, N_i gets slot S_j
15     -- determine the best frame packing configuration β for the TTC
16     for each β_current having a number of 1 to n_β frames do
17       for each frame f_i ∈ β_current do
18         -- determine the best frame size for f_i
19         for each frame size S_f ∈ RecomendedSizes(messages_β) do
20           β_current.f_i.S = S_f
21           -- determine the best frame packing configuration α for the ETC
22           for each α_current having a number of 1 to n_α frames do
23             for each frame f_j ∈ α_current do
24               -- determine the best frame size for f_j
25               for each frame size S_f ∈ RecomendedSizes(messages_α) do
26                 α_current.f_j.S = S_f, ψ_current = <α_current, π_initial, β_current, σ_current>
27                 PackMessages(ψ_current, messages_β ∪ messages_α)
28                 δ_Γ = MultiClusterScheduling(Γ, M, ψ_current)
29                 -- remember the best configuration so far
30                 if δ_Γ(ψ_current) is best so far then ψ_best = ψ_current end if
31               end for
32             end for
33             if ψ_best exists
34             then α_current.f_j.S = size of frame f_j in the configuration ψ_best end if
35           end for
36           if ψ_best exists then α_current = frame set α in the configuration ψ_best end if
37         end for
38       end for
39       if ψ_best exists then β_current.f_i.S = size of frame f_i in the configuration ψ_best end if
40     end for; if ψ_best exists then β_current = frame set β in the configuration ψ_best end if
41   end for
42   if ψ_best exists
43   then σ_current.S_i = node in the slot sequence σ in the configuration ψ_best end if
44 end for
45 return SchedulabilityTest(Γ, ψ_best), ψ_best
end OptimizeFramePacking
```

Figure 10.2: The OptimizeFramePacking Algorithm

The OptimizeFramePacking tries to determine, using the for-each loops in Figure 10.2, the allocation of frames, i.e., the number of frames and their sizes, for each cluster. The actual mapping of messages to frames will be performed by the PackMessages function as described previously.

As an initial TDMA slot sequence $\sigma_{initial}$ on the TTC, OptimizeFramePacking assigns nodes to the slots and fixes the slot length to the minimal allowed value, which is equal to the length of the largest message generated by a process assigned to N_i, $size_{S_i} = size_{largest\_message}$ (line 9 in Figure 10.2).

Then, the algorithm looks, in the innermost for-each loops, for the optimal frame configuration α (lines 21–35). This means deciding on how many frames to include in α (line 22), and which are the best sizes for them (lines 24–31). In α there can be any number of frames, from one single frame to n_α frames (in which case each frame carries one single message). Hence, several numbers of frames are tried, each tested for a recommended size S_f to see if it improves the current configuration. The RecomendedSizes(messages$_\alpha$) list is built recognizing that only messages adjacent in the messages$_\alpha$ list will be packed into the same frame. Sizes of frames are determined as a sum resulted from adding the sizes of combinations of adjacent messages, not exceeding 8 bytes.

> **Example 10.3:** For the previous example, with m_1, m_2 and m_3, of 1, 2 and 3 bytes, respectively, the frame sizes recommended will be of 1, 2, 3, 5, and 6 bytes. A size of 4 bytes will not be recommended since there are no adjacent messages that can be summed together to obtain 4 bytes of data. ■

Once a configuration α_{best} for the ETC, minimizing δ_Γ, has been determined (considering for π, β, σ the initial values determined at the beginning of the algorithm), the algorithm looks for the frame configuration β which will further improve δ_Γ (the loop consisting of lines 15 to 41). The degree of schedulability δ_Γ (the smaller the value, the more schedulable the system) is calculated based on the response times produced by the MultiClusterScheduling algorithm (see Section 8.2) in line 28. After a β_{best} has been decided, the algorithm looks for a slot sequence σ, starting with the first slot and tries to find the node which, when transmitting in this slot, will reduce δ_Γ (loop 11–44). The algorithm continues in this fashion, recording the best ever ψ_{best} configurations obtained in terms of δ_Γ, and thus, the best solution ever is reported when the algorithm finishes. In the inner loops of the heuristic we will not change the frame priorities $\pi_{initial}$ set at the beginning of the algorithm, on line 10.

10.3 Experimental Evaluation

For the evaluation of our algorithms we first used process applications generated for experimental purpose. Similar to the experimental setup in Chapter 8, we considered two-cluster architectures consisting of 2, 4, 6, 8 and 10 nodes, half on the TTC and the other half on the ETC, interconnected by a gateway. Forty processes were assigned to each node, resulting in applications of 80, 160, 240, 320 and 400 processes. Message sizes were randomly chosen between 1 bit and 2 bytes. Thirty examples were generated for each application dimension, thus a total of 150 applications were used for experimental evaluation. Worst-case execution times and message lengths were assigned randomly using both uniform and exponential distribution. For the communication channels we considered a transmission speed of 256 Kbps and a length below 20 meters. All experiments were run on a SUN Ultra 10.

The first result concerns the ability of our heuristics to produce schedulable solutions. We have compared the degree of schedulability δ_Γ obtained from our OptimizeFramePacking (OFP) heuristic (Figure 10.2) with the near-optimal values obtained by the simulated annealing algorithm SA. Obtaining solutions that

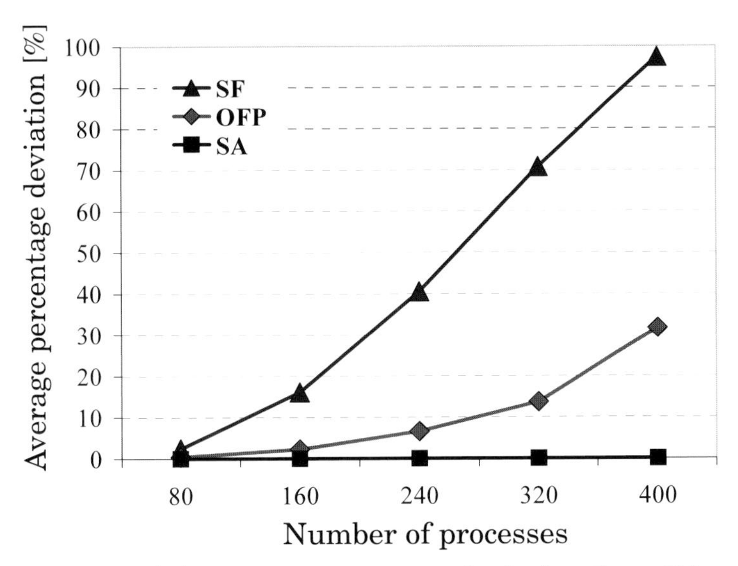

a) Average percentage deviation from SA

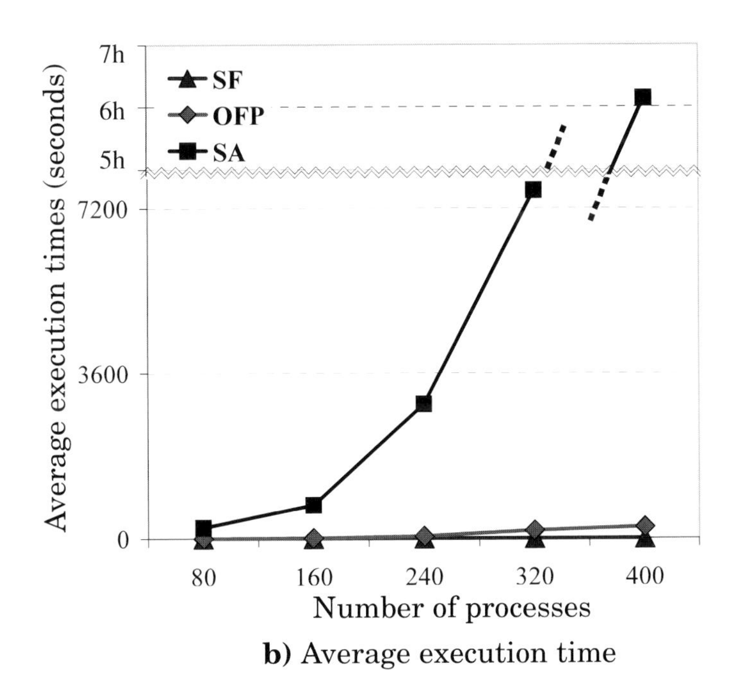

b) Average execution time

Figure 10.3: Evaluation of the Frame Packing Heuristics

have a higher degree of schedulability means obtaining tighter response times, increasing the chances of meeting the deadlines.

Figure 10.3a presents the average percentage deviation of the degree of schedulability produced by OFP from the near-optimal values obtained with SA. Together with OFP, a straightforward approach (SF) is presented. The SF approach does not consider frame packing, and thus each message is transmitted independently in a frame. Moreover, for SF we considered a TTC bus configuration consisting of a straightforward ascending order of allocation of the nodes to the TDMA slots; the slot lengths were selected to accommodate the largest message frame sent by the respective node, and the scheduling has been performed by the MultiClusterScheduling algorithm in Figure 8.2.

Figure 10.3a shows that when packing messages to frames, the degree of schedulability improves dramatically compared to the straightforward approach. The greedy heuristic OptimizeFramePacking performs well for all the graph dimensions, having run-times which are more than two orders of magnitude smaller than with SA.

When deciding on which heuristic to use for design space exploration or system synthesis, an important issue is the execution time. In average, our optimization heuristics needed a couple of minutes to produce results, while the simulated annealing approach had an execution time of up to 6 hours (see Figure 10.3b).

10.3.1 THE VEHICLE CRUISE CONTROLLER

Finally, we considered the cruise controller example presented in Section 2.3.3:

- The model for the cruise controller is presented in Figure 2.9 on page 40.
- The architecture, consisting of a TTC and an ETC, each with 2 nodes, interconnected by the CEM node, is depicted in Figure 2.7b on page 37.

- The software architecture for multi-cluster systems, used by the CC, is presented in Section 3.5.
- We considered one mode of operation with a deadline of 250 ms.

In this context, the straightforward approach SF produced an end-to-end response time of 320 ms, greater than the deadline, while both the OFP and SA heuristics produced a schedulable system with a worst-case response time of 172 ms.

This shows that the optimization heuristic proposed, driven by our schedulability analysis, is able to identify that frame packing configuration which increases the schedulability degree of an application, allowing the developers to reduce the implementation cost of a system.

Appendix A

SIMULATED ANNEALING IS an optimization heuristic that tries to find the global optimum by randomly selecting a new solution from the neighbors of the current solution [Ree93].

The approach derives its name from the process of crystallization of materials. If a material is heated past its melting point and then cooled, the rate of cooling the material can influence its structural properties: a too fast cooling introduces imperfections. This process is called annealing, hence the name *simulated annealing* (SA). Researchers have suggested that this type of simulation can be used to solve optimization problems.

The SA algorithm is a variant of the neighborhood search technique, where the local search space is explored by moving from the current solution to a neighbor solution. In general, the new solution is accepted if it is an improved one. However, in the case of SA, a worse solution can also be accepted with a certain probability that depends on the deterioration of the cost function and on a control parameter called *temperature* which is analog to the temperature concept of the physical annealing process.

In Figure A.1 we give a short description of this algorithm. The algorithm starts with constructing an initial solution. How this initial solution is constructed depends on the particular

```
SimulatedAnnealing
1  construct an initial solution x^now
2  temperature = initial temperature TI
3
4  repeat
5      for i = 1 to temperature length TL do
6          generate randomly a neighboring solution x' of x^now
7          delta = CostFunction(x') – CostFunction(x^now)
8          if delta < 0 then x^now = x'
9          else
10             generate q = random (0, 1)
11             if q < e^(–delta / temperature) then x^now = x' end if
12         end if
13     end for
14     temperature = ε * temperature
15 until stopping criterion is met
16
17 return solution corresponding to the best CostFunction
end SimulatedAnnealing
```

Figure A.1 The Simulated Annealing Strategy

problem that has to be solved. In general, it is sufficient to generate an arbitrary valid solution.

An essential component of the algorithm is the generation of a new solution x' starting from the current one x^{now} (line 5 in the algorithm). The generation of the neighbor solution x' depends on the details of the optimization problem and the internal representation of a solution. The neighbor solutions x' are generated through performing design transformations on x^{now}. The design transformations applied depend on what parts of a system we are interested to synthesize.

For the implementation of the simulated annealing algorithm, the parameters *TI* (initial temperature), *TL* (temperature length), ε (cooling ratio), and the stopping criterion have to be determined. They define the so called cooling schedule and have a decisive impact on the quality of the solutions and the CPU time consumed. The temperature length *TL* determines how many iterations the loop comprised of lines 5–14 will perform at a certain temperature, and the cooling ratio ε will decide how fast the temperature will drop, in each iteration of the 4–15 repeat loop, starting from the initial temperature *TI*.

In our experiments, we were interested to obtain values for *TI*, *TL* and ε that will guarantee the finding of near-optimal solutions in an acceptable time. In order to tune the optimization parameters *TI*, *TL,* and ε we have first performed very long and expensive runs on selected large examples and the best ever solution, for each example, has been considered as the near-optimum. Based on further experiments we have determined the parameters of the SA algorithm, for different sizes of examples, so that the optimization time is reduced as much as possible but the near-optimal result is still produced. These parameters (tuned to different values for each experimental setup) have been then used for the large scale experiments presented in the experimental sections of each chapter.

List of Notations

Application

Γ_i	An application composed of several conditional process graphs
δ_Γ	Degree of schedulability of application Γ
R_Γ	Modification cost of application Γ
Ω	Subset of applications
$R(\Omega)$	Modification cost of the applications in subset Ω
$\mathcal{A}(\mathcal{E}, \mathcal{V})$	Dependency graph of applications; $\mathcal{E}$ is the set of edges, whereas $\mathcal{V}$ is the set of nodes
e_{ij}	An edge $e_{ij} \in \mathcal{E}$ denoting that application Γ_j depends on application Γ_i
S_t	Set of possible worst-case execution times characterizing the execution time of processes belonging to an application
$f_{S_t}(t)$	The occurrence probability of a worst-case execution time $t \in S_t$
S_U	Set of possible worst-case utilizations characterizing the processes of an application
$f_{S_U}(U)$	The occurrence probability of a worst-case utilization $U \in S_U$

S_b — Set of possible message sizes for messages in an application

$f_{S_b}(b)$ — The occurrence probability of a message size $b \in S_b$

T_{min} — Smallest expected period characterizing an application

t_{need} — Expected necessary processor time for an application inside a period T_{min}

b_{need} — Expected necessary bus bandwidth for an application inside a period T_{min}

Conditional Process Graph

$\mathcal{G}_i$ — Conditional process graph $\mathcal{G}_i(V, E_S, E_C)$

V — Set of nodes in the conditional process graph

E_S — Set of simple edges in the conditional process graph

E_S — Set of conditional edges in the conditional process graph

E — The set of all edges in the conditional process graph; $E_S \cup E_C = E$

e_{ij} — Edge in the E set, from P_i to P_j indicating that the output of P_i is the input of P_j

G_i — Mapped conditional process graph

T_{G_i} — Period of the mapped conditional process graph G_i

D_{G_i} — End-to-end deadline on the mapped conditional process graph G_i

τ_i — Process graph without conditions

g_i — A trace through a conditional process graph for a given combination of conditions

Process

P_i	Process
$M(P_i)$	Resource executing process P_i
$\mathcal{N}_{P_i}$	The set of nodes where P_i could potentially be mapped
X_{P_i}	Guard associated to process P_i
C_i	Worst-case execution time of process P_i when executing on the resource $M(P_i)$
U_i	Utilization due to process P_i
T_i	Period of process P_i
$priority_{P_i}$	Priority of process P_i
D_i	Deadline of process P_i
r_i	Worst-case response time of process P_i
J_i	Jitter of process P_i
$hp(P_i)$	Set of processes having a higher priority than $priority_{P_i}$
$lp(P_i)$	Set of processes having a lower priority than $priority_{P_i}$
I_i	The interference of on the execution of process P_i due to processes having a higher priority than $priority_{P_i}$
w_i	The width of the level-i busy period
B_i	The blocking time experienced by process P_i due to processes having a lower priority than $priority_{P_i}$
$ASAP(P_i)$	The as-soon-as-possible start time of process P_i
$ALAP(P_i)$	The as-late-as-possible start time of process P_i
q	Number of level-i busy periods

Message

m	Message
$S(m)$	The sender process of message m
$D(m)$	The destination process of message m
S_m	Size of message m
C_m	Worst-case transmission time of message m
T_m	Period of process message m
$priority_m$	Priority of message m
r_m	Worst-case response time of message m
J_m	Jitter of message m
$hp(m)$	Set of processes having a higher priority than $priority_m$
$lp(m)$	Set of processes having a lower priority than $priority_m$
$I_{m,}\ W_m$	The interference/the worst-case queuing time of message m due to messages having a higher priority than $priority_m$
w_m	The width of the level-m busy period
B_m	The interference on the worst-case transmission time experienced by message m due to processes having a lower priority than $priority_m$
p_m	Number of packets of message m
f	Frame
J_f	Jitter of frame f
p	Packet

System Configuration

ψ	A system configuration
$\mathcal{N}_T$	The TT cluster set of nodes
$\mathcal{N}_E$	The ET cluster set of nodes
N_i	A node in the hardware architecture
N_G	The gateway node
pe_i	Processing element
T_{cycle}	Time length of a TDMA cycle
T_{TDMA}	Time length of a TDMA round
S_i	The ith slot of a TDMA round
S_S	Size of the data field of the largest frame that can be sent in slot S of a TDMA round
θ_m	Maximum time between two consecutive slots of a TDMA cycle carrying message m
s	Transmission speed of a bus
ϕ	Set of offsets
π	Set of priorities
$\mathcal{P}$	Set of processes
α	Set of frames on a CAN bus
β	TDMA round configuration; set of frames on a TTP bus determining the TDMA configuration
σ	Sequence and size of slots in a TDMA round configuration
Out, Out_{N_i}	Queue with messages awaiting transmission on the hardware node N_i
Out_{TTP}	Queue with messages awaiting transmission on the TTP bus from the gateway node of a multi-

	cluster system
Out_{CAN}	Queue with messages awaiting transmission on the CAN bus from the gateway node of a multi-cluster system
s_{out}	Size of an outgoing queue
s_{total}	Total size of all outgoing queues
C	Cost function, design metric

List of Abbreviations

ABS	Anti-lock Braking System
ACK	Acknowledgment
ALAP	As Late As Possible
ASAP	As Soon As Possible
ASIC	Application Specific Integrated Circuit
ASIP	Application Specific Instruction Processor
CAD	Computer Aided Design
CAN	Controller Area Network
CC	Cruise Controller
CEM	Central Electronic Module
CPG	Conditional Process Graph
CPU	Central Processing Unit
CRC	Cyclic Redundancy Check
DSP	Digital Signal Processor
ECM	Engine Control Module
EOF	End Of Field

ET	Event Triggered
ETC	Event Triggered Cluster
ETM	Electronic Throttle Module
FIFO	First In First Out
FPGA	Filed Programmable Gate Array
FPS	Fixed Priority Scheduling
IFD	Inter Frame Delimiter
MBI	Message Base Interface
MDEL	Message Descriptor List
MHTT	Message Handling Time Table
MPCP	Modified Partial Critical Path
PCP	Partial Critical Path
RAM	Random Access Memory
ROM	Read Only Memory
SA	Simulated Annealing
SOF	Start Of Frame
TCM	Transmission Control Module
TDMA	Time Division Multiple Access
TT	Time Triggered
TTC	Time Triggered Cluster
TTP	Time Triggered Protocol
VLSI	Very Large Scale Integration
WCAO	Worst Case Administrative Overhead
WCET	Worst Case Execution Time

Index

Bibliography

[Aar03] E. Aarts, "IC Design Challenges for Ambient Intelligence," in *Proceedings of the Design Automation and Test in Europe Conference*, pages 2–7, 2003.

[Agr94] G. Agrawal, B. Chen, W. Zhao, S. Davari, "Guaranteeing Synchronous Message Deadlines with the Token Medium Access Control Protocol," in *IEEE Transactions on Computers*, volume 43, issue 3, pages 327–339, March 1994.

[Aud91] N. C. Audsley, A. Burns, M. F. Richardson, A. J. Wellings, "Hard Real-Time Scheduling: The Deadline Monotonic Approach," in *Proceedings of the 8th IEEE Workshop on Real-Time Operating Systems and* Software, pages 127–132, 1991.

[Aud93] N. C. Audsley, K. Tindell, A. Burns, "The End Of The Line For Static Cyclic Scheduling?," in *Proceedings of the 5th Euromicro Workshop on Real-Time Systems*, 36–41, 1993.

[Aud95] N. C. Audsley, A. Burns, R. I. Davis, K. W. Tindell, A. J. Wellings, "Fixed Priority Pre-emptive Scheduling: An Historical Perspective," in *Real-Time Systems*, volume 8, pages 173–198, 1995.

[Axe96] J. Axelsson, "Hardware/Software Partitioning Aiming at Fulfilment of Real-Time Constraints," in *Journal of Systems Architecture*, volume 42, issues 6–7, pages 449–464, December 1996.

[Bal97] F. Balarin, editor, *Hardware-Software Co-Design of Embedded Systems: The Polis Approach*, Kluwer Academic Publishers, 1997.

[Bal98] F. Balarin, L. Lavagno, P. Murthy, A. Sangiovanni-Vincentelli, "Scheduling for Embedded Real-Time Systems," in *IEEE Design & Test of Computers*, volume 15, issue 1, pages 71–82, January–March 1998.

[Bar98a] S. Baruah, "Feasibility Analysis of Recurring Branching Tasks," in *Proceedings of the 10th Euromicro Workshop on Real-Time Systems*, pages 138–145, 1998.

[Bar98b] S. Baruah, "A General Model for Recurring Real-Time Tasks," in *Proceedings of the IEEE Real-Time Symposium*, pages 114–122, 1998.

[Bec98] J. E. Beck, D. P. Siewiorek, "Automatic Configuration of Embedded Multicomputer Systems," in *IEEE Transactions on CAD*, volume 17, number 2, pages 84–95, 1998.

[Ber00] J. Berwanger, M. Peller, R. Griessbach, *A New High Performance Data Bus System for Safety-Related Applications*, http://www.byteflight.de, 2000.

[Ben96] A. Bender, "Design of an Optimal Loosely Coupled Heterogeneous Multiprocessor System," in *Proceedings of the Electronic Design and Test Conference*, pages 275–281, 1996.

[Bin01] E. Bini, G. Butazzo, G. Butazzo, "A Hyperbolic Bound for the Rate Monotonic Algorithm," in *Proceedings of the 13th Euromicro Conference on Real-Time Systems*, pages 59–66, 2001.

[Boe00] B. W. Boehm et al., *Software Cost Estimation with COCOMO II*, Prentice-Hall, 2000.

[Bol97] I. Bolsens, H. J. De Man, B. Lin, K. Van Rompaey, S. Vercauteren, D. Verkest, "Hardware/Software Co-Design of Digital Telecommunication Systems," in *Proceedings of the IEEE*, volume 85, issue 3, pages 391–418, 1997.

[Bos91] R. Bosch GmbH, *CAN Specification Version 2.0*, 1991.

[Chi96] M. Chiodo, "Automotive Electronics: A Major Application Field for Hardware-Software Co-Design," in *Hardware/Software Co-Design,* Kluwer Academic Publishers, pages 295–310, 1996.

[Cho92] P. Chou, R. Ortega, G. Borriello, "Synthesis of Hardware/Software Interface in Microcontroller-Based Systems," in *Proceedings of the International Conference on Computer Aided Design*, pages 488–495, 1992.

[Cho95a] P. H. Chou, R. B. Ortega, G. Borriello, "The Chinook Hardware/Software Co-Synthesis System," in *Proceedings of the International Symposium on System Synthesis*, pages 22–27, 1995.

[Cho95b] P. Chou, G. Borriello, "Interval Scheduling: Fine-Grained Code Scheduling for Embedded Systems," in *Proceedings of the Design Automation and Test Conference*, pages 462–467, 1995.

[Cof72] E. G. Coffman Jr., R. L. Graham, "Optimal Scheduling for two Processor Systems," in *Acta Informatica*, issue 1, pages 200–213, 1972.

[Dav95] J. M. Daveau, T. Ben Ismail, A. A. Jerraya, "Synthesis of System-Level Communication by an Allocation-Based Approach," in *Proceedings of the International Symposium on System Synthesis*, pages 150–155, 1995.

[Dav98] B. P. Dave, N. K. Jha, "COHRA: Hardware-Software Cosynthesis of Hierarchical Heterogeneous Distributed Systems," in *IEEE Transactions on CAD*, volume 17, number 10, pages 900–919, 1998

[Dav99] B. P. Dave, G. Lakshminarayana, N. J. Jha, "COSYN: Hardware-Software Co-Synthesis of Heterogeneous Distributed Embedded Systems," in *IEEE Transactions on VLSI Systems*, volume 7, number 1, pages 92–104, 1999.

[Deb97] J. A. Debardelaben, V. K. Madiseti, A. J. Gadient, "Incorporating Cost Modeling in Embedded-System Design," in *IEEE Design & Test of Computers*, volume 14, number 3, pages 24–35, July–September 1997.

[Deo98] J. S. Deogun, R. M. Kieckhafer, A. W. Krings, "Stability and Performance of List Scheduling with External Process Delays," in *Real Time Systems*, volume 15, number 1, pages 5–38, 1998.

[Dic98] R. P. Dick, N. K. Jha, "CORDS: Hardware-Software Co-Synthesis of Reconfigurable Real-Time Distributed Embedded Systems," in *Proceedings of the International Conference on CAD*, pages 62–67, 1998.

[Dob98] A. Doboli, P. Eles, "Scheduling under Control Dependencies for Heterogeneous Architectures," in *International Conference on Computer Design*, pages 602–608, 1998.

[Eas02] EAST-EEA project, *ITEA Full Project Proposal,* http://www.itea-office.org, 2002.

[EB03a] Encyclopædia Britannica Online, *Computers,* http://search.eb.com/eb/article?eu=130080, 2003.

[EB03b] Encyclopædia BritannicaOnline, *Intel Corporation,* http://search.eb.com/eb/article?eu=2242, 2003.

[Ech03] Echelon, *LonWorks: The LonTalk Protocol Specification*, http://www.echelon.com, 2003.

[Edw97] S. Edwards, L. Lavagno, E. A. Lee, A.Sangoivanni-Vincentelli, "Design of Embedded Systems: Formal Models, Validation and Synthesis," in *Proceedings of the IEEE*, volume 85, issue 3, pages 366–390, March 1997.

[Edw00] S. Edwards, *Languages for Digital Embedded Systems,* Kluwer Academic Publishers, 2000.

[Ele97] P. Eles, Z. Peng, K. Kuchcinski, A. Doboli, "System Level Hardware/Software Partitioning Based on Simulated Annealing and Tabu Search," in *Design Automation for Embedded Systems*, volume 2, number 1, pages 5–32, 1997.

[Ele98a] P. Eles, K. Kuchcinski, Z. Peng, A. Doboli, P. Pop, "Scheduling of Conditional Process Graphs for the Synthesis of Embedded Systems," in *Proceedings of Design Automation and Test in Europe*, pages 132–139, 1998.

[Ele98b] P. Eles, K. Kuchcinski, Z. Peng, A. Doboli, P. Pop, "Process Scheduling for Performance Estimation and Synthesis of Hardware/Software Systems," in *Proceedings of the Euromicro Conference*, pages 168–175, 1998.

[Ele00] P. Eles, A. Doboli, P. Pop, Z. Peng, "Scheduling with Bus Access Optimization for Distributed Embedded Systems," in *IEEE Transactions on VLSI Systems,* volume 8, number 5, pages 472–491, 2000.

[Ele02] P. Eles, *Lecture Notes for System Design and Methodology*, http://www.ida.liu.se/~TDTS30, 2002.

[Eng99] J. Engblom, A. Ermedahl, M. Sjödin, J. Gustafsson, H. Hansson, "Towards Industry Strength Worst-Case Execution Time Analysis," in *Swedish National Real-Time Conference*, 1999.

[Erm97] H. Ermedahl, H. Hansson, M. Sjödin, "Response-Time Guarantees in ATM Networks," in *Proceedings of the IEEE Real-Time Systems Symposium*, pages 274–284, 1997.

[Ern93] R. Ernst, J. Henkel, T. Benner, "Hardware/Software Co-synthesis for Microcontrollers," in *IEEE Design & Test of Computers*, pages 64–75, September 1997.

[Ern97] R. Ernst, W. Ye, "Embedded Program Timing Analysis Based on Path Clustering and Architecture Classification," in *Proceedings of the International Conference on CAD*, pages 598–604, 1997.

[Ern98] R. Ernst, "Codesign of Embedded Systems: Status and Trends," in *IEEE Design and Test of Computers*, volume 15, number 2, pages 45–54, April–June 1998.

[Thi99] L. Thiele, K. Strehl, D. Ziegengein, R. Ernst, J. Teich, "FunState—An Internal Design Representation for Codesign," in *International Conference on Computer-Aided Design*, pages 558–565, 1999.

[Fle02] *FlexRay Requirements Specification*, http://www.flexray-group.com/, 2002.

[Foh93] G. Fohler, "Realizing Changes of Operational Modes with Pre Run-time Scheduled Hard Real-Time Systems," in *Responsive Computer Systems,* H. Kopetz and Y. Kakuda, editors, pages 287–300, Springer Verlag, 1993.

[Gaj83] D. D. Gajski, R. H. Kuhn, "Guest Editor's Introduction: New VLSI Tools," in *IEEE Computer*, December 1983.

[Gaj95] D. D. Gajski, F. Vahid, "Specification and Design of Embedded Hardware-Software Systems," in *IEEE Design and Test of Computers*, volume 12, number 1, pages 53–67, Spring 1995.

[Ger96] R. Gerber, D. Kang, S. Hong, M. Saksena, "End-to-End Design of Real-Time Systems," in *Formal Methods in Real-Time Computing*, D. Mandrioli and C. Heitmeyer, editors, John Wiley & Sons, 1996.

[Gon95] J. Gong, D. D. Gajski, S. Narayan, "Software Estimation Using A Generic-Processor Model," in *Proceedings of the European Design and Test Conference*, pages 498–502, 1995.

[Gup93] R. K. Gupta, G. De Micheli, “Hardware-Software Cosynthesis for Digital Systems,” in *IEEE Design & Test of Computers*, volume 10, number 3, pages 29–41, September 1993.

[Gup95] R. K. Gupta, *Co-Synthesis of Hardware and Software for Digital Embedded Systems*, Kluwer Academic Publishers, Boston, 1995.

[Gut95] J, J, Gutiérrez García, M. González Harbour, “Optimized Priority Assignment for Tasks and Messages in Distributed Hard Real-Time Systems,” in *Proceedings of the 3rd Workshop on Parallel and Distributed Real-Time Systems*, pages 124–132, 1995.

[Han02] P. Hansen, *The Hansen Report on Automotive Electronics*, http://www.hansenreport.com/, July–August, 2002.

[Hau02] C. Haubelt, J. Teich, K. Richter, R. Ernst, “System Design for Flexibility,” in *Proceedings of the Design, Automation and Test in Europe Conference*, pages 854–861, 2002.

[Hen95] J. Henkel, R. Ernst, “A Path-Based Technique for Estimating Hardware Run-time in Hardware/Software Cosynthesis,” in *Proceedings of the International Symposium on System Synthesis*, pages 116–121, 1995.

[Hoy92] K. Hoyme, K. Driscoll, “SAFEbus,” in *IEEE Aerospace and Electronic Systems Magazine*, volume 8, number 3, pages 34–39, 1992.

[Int02] International Organization for Standardization, “Road vehicles—Controller area network (CAN)—Part 4: Time-triggered communication”, ISO/DIS 11898-4, 2002.

[Jan02] J. Jonsson, K. G. Shin, "Robust Adaptive Metrics for Deadline Assignment in Distributed Hard Real-Time Systems," *Real-Time Systems: The International Journal of Time-Critical Computing Systems*, Vol. 23, No. 3, pages 239–271, 2002.

[Jor97] P. B. Jorgensen, J. Madsen, "Critical Path Driven Cosynthesis for Heterogeneous Target Architectures," in *Proceedings of the International Workshop on Hardware / Software Codesign*, pages 15–19, 1997.

[Jos01] K. Jost, "From Fly-by-Wire to Drive-by-Wire," *Automotive Engineering International*, 2001.

[Kal97] A. Kalawade, E.A. Lee, "The Extended Partitioning Problem: Hardware/Software Mapping, Scheduling, and Implementation-Bin Selection," in *Design Automation for Embedded Systems*, volume 2, pages 125–163, 1997.

[Kas84] H. Kasahara, S. Narita, "Practical Multiprocessor Scheduling Algorithms for Efficient Parallel Processing," in *IEEE Transaction on Computers*, volume 33, number 11, pages 1023–1029, 1984.

[Kie97] B. Kienhuis, E. Deprettere, K. Vissers, P. Van Der Wolf, "An Approach for Quantitative Analysis of Application-Specific Dataflow Architectures," in *Proceedings of the IEEE International Conference on Application-Specific Systems, Architectures and Processors*, pages 338 –349, 1997.

[Keu00] K. Keutzer, S. Malik, A. R. Newton, "System-Level Design: Orthogonalization of Concerns and Platform-Based Design," in *IEEE Transactions on Computer-Aided Design of Integrated Circuits and Systems*, volume 19, number 12, December 2000.

[Kla01] S. Klaus, S. A. Huss, "Interrelation of Specification Method and Scheduling Results in Embedded System Design," in *Proceedings of the ECSI International Forum on Design Languages*, 2001.

[Knu99] P. V. Knudsen, J. Madsen, "Integrating Communication Protocol Selection with Hardware/Software Codesign," in *IEEE Transactions on CAD*, volume 18, number 8, pages 1077–1095, 1999.

[Kop95] H. Kopez, R. Nossal, "The Cluster-Compiler—A Tool for the Design of Time Triggered Real-Time Systems," in *Proceedings of the ACM SIGPLAN Workshop. on Languages, Compilers, and Tools for Real-Time Systems*, pages 108–116, 1995.

[Kop97a] H. Kopetz, *Real-Time Systems-Design Principles for Distributed Embedded Applications*, Kluwer Academic Publishers, 1997.

[Kop97b] H. Kopetz et al., "A Prototype Implementation of a TTP/C Controller," in *SAE Congress and Exhibition*, 1997.

[Kop99] H. Kopetz, "Automotive Electronics," in *Proceedings of the 11th Euromicro Conference on Real-Time Systems*, pages 132–140, 1999.

[Kop01] H. Kopetz, *A Comparison of TTP/C and FlexRay*, technical report 2001/10, Technical University Vienna, 2001.

[Kop03] H. Kopetz, G. Bauer, "The Time-Triggered Architecture," in *Proceedings of the IEEE*, volume 91, issue 1, pages 112–126, 2003.

[Kuc97] K. Kuchcinski, "Embedded System Synthesis by Timing Constraint Solving," in *Proceedings of the International Symposium on System Synthesis*, pages 50–57, 1997.

[Kuc01] K. Kuchcinski, “Constraints Driven Design Space Exploration for Distributed Embedded Systems,” in *Journal of Systems Architecture*, volume 47, issues 3–4, pages 241–261, 2001.

[Kwo96] Y. K. Kwok, I. Ahmad, “Dynamic Critical-Path Scheduling: an Effective Technique for Allocating Task Graphs to Multiprocessors,” in *IEEE Transactions on Parallel and Distributed Systems*, volume 7, number 5, pages 506–521, 1996.

[Lak99] G. Lakshminarayana, K. S. Khouri, N. K. Jha, “Wavesched: A Novel Scheduling Technique for Control-Flow Intensive Designs,” in *IEEE Transactions on Computer-Aided Design of Integrated Circuits and Systems,* volume 18, number 5, pages 108–113, 1999.

[Lav99] L. Lavagno, A. Sangiovanni-Vincentelli, and E. Sentovich, “Models of Computation for Embedded System Design,” in *System-Level Synthesis*, Kluwer Academic Publishers, pages 45–102, 1999.

[Lee95] E. A. Lee, T. M. Parks, “Dataflow process networks,” in *Proceedings of the IEEE*, volume 83, pages 773–801, May 1995.

[Lee99] C. Lee, M. Potkonjak, W. Wolf, “Synthesis of Hard Real-Time Application Specific Systems,” in *Design Automation for Embedded Systems*, volume 4, issue 4, pages 215–241, 1999.

[Lee02] G. Leen, D. Hefffernan, “Expanding Automotive Electronic Systems,” in *IEEE Computer*, pages 88–93, January 2002.

[Leh89] J. Lehoczky, L. Sha, Y. Ding, “The Rate Monotonic Scheduling Algorithm: Exact Characterization and Average Case Behaviour,” in *Proceedings of the 11th Real-Time Symposium*, pages 166–171, 1989.

[Leh90] J. P. Lehoczky, "Fixed Priority Scheduling of Periodic Task Sets With Arbitrary Deadlines," in *Proceedings of 11th IEEE Real-Time Symposium*, pages 201–209, 1990.

[Li95] Y. S. Li, S. Malik, "Performance Analysis of Embedded Software Using Implicit Path Enumeration," in *Proceedings of the Design Automation Conference*, pages 456–461, 1995.

[Li00] Y. Li, T. Callahan, E. Darnell, R. Harr, U. Kurkure, J. Stockwood, "Hardware-Software Co-Design of Embedded Reconfigurable Architectures," in *Proceedings of the Design Automation Conference*, pages 507–512, 2000.

[Lie99] P. Lieverse, P. van der Wolf, E. Deprettere, K. Vissers, "A Methodology for Architecture Exploration of Heterogeneous Signal Processing Systems," in *IEEE Workshop on Signal Processing Systems*, pages 181–190, 1999.

[Liu73] C. L. Liu, J. W. Layland, "Scheduling Algorithms for Multiprogramming in a Hard Real-Time Environment," in *Journal of the ACM*, volume 20, number 1, pages 46–61, 1973.

[Lön99] H. Lönn, J. Axelsson, "A Comparison of Fixed-Priority and Static Cyclic Scheduling for Distributed Automotive Control Applications," in *Proceedings of the 11th Euromicro Conference on Real-Time Systems*, pages 142–149, 1999.

[Lun99] T. Lundqvist, P. Stenström, "An Integrated Path and Timing Analysis Method Based on Cycle-Level Symbolic Execution," in *Real-Time Systems*, volume 17, number 2–3, pages 183–207, 1999.

[Mal97] S. Malik, M. Martonosi, Y.S. Li, "Static Timing Analysis of Embedded Software," in *Proceedings of the Design Automation Conference*, pages 147–152, 1997.

[Mar90] S. Martello, P. Toth, *Kanpsack Problems: Algorithms and Computer Implementations*, Wiley, 1990.

[Mar02] G. Martin, F. Schirrmeister, "A Design Chain for Embedded Systems," *Computer*, volume 35, issue 3, pages 100–103, March 2002.

[Mat03] Matlab/Simulink, http://www.mathworks.com

[Mc92] K. McMillan, D. Dill, "Algorithms for Interface Timing Verification," in *Proceedings of the IEEE International Conference on Computer Design*, pages 48–51, 1992.

[Mic96] G. De Micheli, M.G. Sami, editors, "Hardware/Software Co-Design," *NATO ASI 1995*, Kluwer Academic Publishers, 1996.

[Mic97] G. De Micheli, R.K. Gupta, "Hardware/Software Co-Design," in *Proceedings of the IEEE*, volume 85, issue 3, pages 349–365, 1997.

[Min00] P. S. Miner, "Analysis of the SPIDER Fault-Tolerance Protocols," in *Proceedings of the 5th NASA Langley Formal Methods Workshop*, 2000.

[Moo97] V. Mooney, T. Sakamoto, G. De Micheli, "Run-Time Scheduler Synthesis for Hardware-Software Systems and Application to Robot Control Design," in *Proceedings of the International Workshop on Hardware-Software Co-design*, pages 95–99, 1997.

[Nar94] S. Narayan, D. D. Gajski, "Synthesis of System-Level Bus Interfaces," in *Proceedings of the European Design and Test Conference*, pages 395–399, 1994.

[Nol01] T. Nolte, H. Hansson, C. Norström, S. Punnekkat, "Using bit-stuffing distributions in CAN analysis," in *Proceedings of the IEEE/IEE Real-Time Embedded Systems Workshop*, 2001

[Ort98] R. B. Ortega, G. Borriello, "Communication Synthesis for Distributed Embedded Systems," in *Proceedings of the International Conference on CAD*, pages 437–444, 1998.

[Pal97] J. C. Palencia, J. J. Gutiérrez García, M. González Harbour, "On the Schedulability Analysis for Distributed Hard Real-Time Systems," in *Proceedings of the Euromicro Conference on Real Time Systems*, pages 136–143, 1997.

[Pal98] J. C. Palencia, M. González Harbour, "Schedulability Analysis for Tasks with Static and Dynamic Offsets," in *Proceedings of the 19th IEEE Real-Time Systems Symposium*, pages 26–37, 1998.

[Pal99] J. C. Palencia, M. González Harbour, "Exploiting Precedence Relations in the Schedulability Analysis of Distributed Real-Time Systems," in *Proceedings of the 20th IEEE Real-Time Systems Symposium*, pages 328–339, 1999.

[Pop98] P. Pop, P. Eles, Z. Peng, "Scheduling Driven Partitioning of Heterogeneous Embedded Systems," in *Proceedings of the Swedish Workshop on Computer Systems Architecture,* pages 99–102, 1998.

[Pop99a] P. Pop, P. Eles, Z. Peng, "Scheduling with Optimized Communication for Time-Triggered Embedded Systems," in *Proceedings of the 7th International Workshop on Hardware/Software Codesign*, pages 178–182, 1999.

[Pop99b] P. Pop, P. Eles, Z. Peng, "Communication Scheduling for Time-Triggered Systems," in *Work in Progress Proceedings of the 11th Euromicro Conference on Real-Time Systems,* 1999.

[Pop99c] P. Pop, P. Eles, Z. Peng, "An Improved Scheduling Technique for Time-Triggered Embedded Systems," in *Proceedings of the 25th Euromicro Conference,* pages 303–310, 1999.

[Pop99d] P. Pop, P. Eles, Z. Peng, "Schedulability-Driven Communication Synthesis for Time Triggered Embedded Systems," in *Proceedings of the 6th International Conference on Real-Time Computing Systems and Applications,* pages 287–294, 1999.

[Pop00a] P. Pop, P. Eles, Z. Peng, "Bus Access Optimization for Distributed Embedded Systems Based on Schedulability Analysis," in *Proceedings of the Design, Automation and Test in Europe Conference,* pages 567–574, 2000.

[Pop00b] P. Pop, P. Eles, Z. Peng, "Performance Estimation for Embedded Systems with Data and Control Dependencies," in *Proceedings of the 8th International Workshop on Hardware/Software Codesign,* pages 62–66, 2000.

[Pop00c] P. Pop, P. Eles, Z. Peng, "Schedulability Analysis for Systems with Data and Control Dependencies," in *Proceedings of the 12th Euromicro Conference on Real-Time Systems*, pages 201–208, 2000.

[Pop01a] P. Pop, P. Eles, T. Pop, Z. Peng, "An Approach to Incremental Design of Distributed Embedded Systems," in *Proceedings of the Design Automation Conference*, pages 450–455, 2001.

[Pop01b] P. Pop, P. Eles, T. Pop, Z. Peng, "Minimizing System Modification in an Incremental Design Approach," in *Proceedings of the International Workshop on Hardware/Software Codesign*, pages 183–188, 2001.

[Pop02a] P. Pop, P. Eles, Z. Peng, "Flexibility Driven Scheduling and Mapping for Distributed Real-Time Systems," in *Proceedings of the International Conference on Real-Time Computing Systems and Applications*, pages 337–346, 2002.

[Pop02b] T. Pop, P. Eles, Z. Peng, "Holistic Scheduling and Analysis of Mixed Time/Event-Triggered Distributed Embedded Systems," in *International Symposium on Hardware/Software Codesign*, pages 187–192, 2002.

[Pop02c] T. Pop, P. Eles, Z. Peng, "Design Optimization of Mixed Time/Event-Triggered Distributed Embedded Systems", in *CODES+ISSS (merged conference)*, pages 83–89, 2003.

[Pop02d] P. Pop, P. Eles, Z. Peng, "Schedulability-Driven Communication Synthesis for Time-Triggered Embedded Systems," to be published in *Journal of Real-Time Systems*.

[Pop03a] P. Pop, P. Eles, Z. Peng, "Schedulability Analysis and Optimization for the Synthesis of Multi-Cluster Distributed Embedded Systems," in *Proceedings of the Design Automation and Test in Europe Conference*, pages 184–189, 2003.

[Pop03b] P. Pop, P. Eles, Z. Peng, "Schedulability-Driven Frame Packing for Multi-Cluster Distributed Embedded Systems," in *Proceedings of the ACM SIGPLAN Conference on Languages, Compilers and Tools for Embedded Systems*, pages 113–122, 2003.

[Pop04] P. Pop, P. Eles, Z. Peng, "Design Optimization of Multi-Cluster Embedded Systems for Real-Time Applications", in *Proceedings of the Design, Automation and Test in Europe Conference*, 2004.

[Pra92] S. Prakash, A. Parker, "SOS: Synthesis of Application-Specific Heterogeneous Multiprocessor Systems," in *Journal of Parallel and Distributed Computers*, volume 16, pages 338–351, 1992.

[Pro03] Profibus International, *PROFIBUS DP Specification*, http://www.profibus.com/, 2003.

[Raj98] A. Rajnak, K. Tindell, L. Casparsson, *Volcano Communications Concept*, Volcano Communication Technologies AB, 1998.

[Rag02] D. Ragan, P. Sandborn, P. Stoaks, "A Detailed Cost Model for Concurrent Use with Hardware/Software Co-Design," in *Proceedings of the Design Automation Conference*, pages 269–274, 2002.

[Ree93] C. R. Reevs, *Modern Heuristic Techniques for Combinatorial Problems*, Blackwell Scientific Publications, 1993.

[REV94] *REVIC Software Cost Estimating Model*, User's Manual, V9.0–9.2, US Air Force Analysis Agency, 1994.

[Ric02] K. Richter, R. Ernst, "Event Model Interfaces for Heterogeneous System Analysis," in *Proceedings of the Design Automation and Test in Europe Conference*, pages 506–513, 2002.

[Ric03] K. Richter, M. Jersak, R. Ernst, "A Formal Approach to MpSoC Performance Verification," in *Computer*, volume 36, issue 4, pages 60–67, 2003.

[Rho99] D. L. Rhodes, Wayne Wolf, "Co-Synthesis of Heterogeneous Multiprocessor Systems using Arbitrated Communication," in *Proceeding of the International Conference on CAD*, pages 339–342, 1999.

[Rus01] J. Rushby, "Bus Architectures for Safety-Critical Embedded Systems," *Springer–Verlag Lecture Notes in Computer Science*, volume 2211, pages 306–323, 2001.

[San00] K. Sandström, C. Norström, "Frame Packing in Real-Time Communication," in *Proceedings of the International Conference on Real-Time Computing Systems and Applications*, pages 399–403, 2000.

[San03] A. Sangiovanni-Vincentelli, "Electronic-System Design in the Automobile Industry", in *IEEE Micro*, volume 23, issue 3, pages 8–18, 2003.

[Sha90] L. Sha, R. Rajkumar, J. Lehoczky, "Priority Inheritance Protocols: An Approach to Real-Time Synchronization," in *IEEE Transactions on Computers*, volume 39, number 9, pages 1175–1185, 1990.

[Sem02] Semiconductor Industry Association, *The International Technology Roadmap for Semiconductors*, http://public.itrs.net/, 2002.

[Sof97] *SoftEST—Version 1.1*, US Air Force Analysis Agency, 1997.

[Sta97] J. Staunstrup, W. Wolf, editors, *Hardware/Software Co-Design: Principles and Practice*, Kluwer Academic Publishers, 1997.

[Str89] J. K. Strosnider, T. E. Marchok, "Responsive, Deterministic IEEE 802.5 Token Ring Scheduling," in *Journal of Real-Time Systems*, volume 1, issue 2, pages 133–158, 1989.

[Suz96] K. Suzuki, A. Sangiovanni-Vincentelli, "Efficient Software Performance Estimation Methods for Hardware/Software Codesign," in *Proceedings of the Design Automation Conference*, pages 605–610, 1996.

[Sta93] J. A. Stankovic, K. Ramamritham, editors, *Advances in Real-Time Systems*, IEEE Computer Society Press, 1993.

[Tab00] B. Tabbara, A. Tabbara, A. Sangiovanni-Vincentelli, *Function/Architecture Optimization and Co-Design of Embedded Systems*, Kluwer Academic Publishers, 2000.

[Tec02] TTTech, *Comparison CAN–Byteflight–FlexRay–TTP/C*, Technical Report, availabe at http://www.tttech.com/

[Tin92] K. Tindell, A. Burns, A. J. Wellings, "Allocating Real-Time Tasks (An NP-Hard Problem made Easy)," in *Journal of Real-Time Systems*, volume 4, issue 2, pages 145–165, 1992.

[Tin94a] K. Tindell, J. Clark, "Holistic Schedulability Analysis for Distributed Hard Real-Time Systems," in *Microprocessing and Microprogramming*, volume 40, pages 117–134, 1994.

[Tin94b] K. Tindell, *Adding Time-Offsets to Schedulability Analysis*, Technical Report Number YCS–94–221, Department of Computer Science, University of York, 1994.

[Tin95] K. Tindell, A. Burns, A. J. Wellings, "Calculating Controller Area Network (CAN) Message Response Times," in *Control Engineering Practice*, volume 3, number 8, pages 1163–1169, 1995.

[Tur99] J. Turley, "Embedded Processors by the Numbers," in *Embedded Systems Programming*, May 1999.

[Ull75] D. Ullman, "NP-Complete Scheduling Problems," in *Journal of Computer Systems Science*, volume 10, pages 384–393, 1975.

[Vah94] F. Vahid, J. Gong, D. Gajski, "A Binary-Constraint Search Algorithm for Minimizing Hardware during Hardware/Software Partitioning," in *Proceedings of the European Design Automation Conference*, pages 214–219, 1994.

[Vah02] F. Vahid, T. Givargis, *Embedded Systems Design: A Unified Hardware/Software Introduction*, John Wiley & Sons, 2002.

[Val95] C. A. Valderrama, A. Changuel, P. V. Raghavan, M. Abid, T. Ben Ismail, A. A. Jerraya, "A Unified Model for Co-Simulation and Co-Synthesis of Mixed Hardware/Software Systems," in *Proceedings of the European Design and Test Conference*, pages 180–184, 1995.

[Val96] C. A. Valderrama, F. Nacabal, P. Paulin, A. A. Jerraya, "Automatic Generation of Interfaces for Distributed C-VHDL Cosimulation of Embedded Systems: an Industrial Experience," in *Proceedings of the International Workshop on Rapid System Prototyping*, pages 72–77, 1996.

[Ver96] D. Verkest, K. Van Rompaey, I. Bolsens, H. De Man, "CoWare—A Design Environment for Heterogeneous Hardware/Software Systems," in *Design Automation for Embedded Systems,* volume 1, pages 357–386, 1996.

[Wal94] E. Walkup, G. Borriello, *Automatic Synthesis of Device Drivers for Hardware/Software Co-design*, Technical Report 94–06–04, Dept. of Computer Science and Engineering, University of Washington, 1994.

[Wol01] W. Wolf, *Computers as Components: Principles of Embedded Computing System Design*, Morgan Kaufmann Publishers, 2001.

[Wol02] F. Wolf, J. Staschulat, R. Ernst, "Associative Caches in Formal Software Timing Analysis," in *Proceedings of the 39th Conference on Design Automation*, pages 622–627, 2002.

[Wol94] W. Wolf, "Hardware-Software Co-Design of Embedded Systems," in *Proceedings of the IEEE*, volume 82, number 7, pages 967–989, 1994.

[Wol97] W. Wolf, "An Architectural Co-Synthesis Algorithm for Distributed, Embedded Computing Systems," in *IEEE Transactions on VLSI Systems*, volume 5, number 2, pages 218–229, June 1997.

[Wol03] W. Wolf, "A Decade of Hardware/Software Codesign," in *Computer*, volume 36, issue 4, pages 38–43, 2003.

[Wu90] M. Y. Wu, D. D. Gajski, "Hypertool: A Programming Aid for Message-Passing Systems," in *IEEE Transactions on Parallel and Distributed Systems*, volume 1, number 3, pages 330–343, 1990.

[XbW98] X-by-Wire Consortium, *X-By-Wire: Safety Related Fault Tolerant Systems in Vehicles*, http://www.vmars.tuwien.ac.at/projects/xbywire/, 1998.

[Xie01] Y. Xie, W. Wolf, "Allocation and Scheduling of Conditional Task Graph in Hardware/Software Co-Synthesis," in *Proceedings of the Design Automation and Test in Europe Conference*, pages 620–625, 2001.

[Xu00] J. Xu, D. L. Parnas, "Priority Scheduling Versus Pre-Run-Time Scheduling," in *Journal of Real Time Systems*, volume 18, issue 1, pages 7–24, 2000.

[Xu93] J. Xu, D. L. Parnas, "On Satisfying Timing Constraints in Hard-Real-Rime Systems," in *IEEE Transactions Software Engineering*, volume 19, number 1, pages 70–84, 1993.

[Yen97] T. Y. Yen, W. Wolf, *Hardware-Software Co-Synthesis of Distributed Embedded Systems*, Kluwer Academic Publishers, 1997.

[Yen98] T. Yen, W. Wolf, "Performance Estimation for Real-Time Distributed Embedded Systems," in *IEEE Transactions on Parallel and Distributed Systems*, volume 9, number 11, pages 1125–1136, 1998.